P9-DCX-441

ALSO BY T. R. REID

Congressional Odyssey

The Pursuit of the Presidency, 1980 (coauthor)

The Chip

How Two Americans Invented the Microchip and Launched a Revolution

by T.R. Reid

SIMON AND SCHUSTER · NEW YORK

Published by Simon and Schuster
A Division of Simon & Schuster, Inc.
Simon & Schuster Building
Rockefeller Center
1230 Avenue of the Americas
New York, New York 10020
SIMON AND SCHUSTER and colophon are registered trademarks
of Simon & Schuster, Inc.
Designed by Eve Kirch
Manufactured in the United States of America

10 9 8 7 6 5 4 3 2 1

Library of Congress Cataloging in Publication Data

Reid, T. R.
The chip: how two Americans invented the microchip
and launched a revolution.
Bibliography: p.
Includes index.
1. Microelectronics—History. I. Title.
TK7874.R43 1984 621.381'7'09 84-22148

ISBN: 0-671-45393-9

Matri filioque in amore dedicatus.

Contents

1

The Tyranny of Numbers

THE IDEA OCCURRED to Jack Kilby at the height of summer, when everyone else was on vacation and he had the lab to himself. It was an idea, as events would prove, of literally cosmic dimensions, an idea that would change the daily life of the world and be honored in the textbooks with a name of its own: The Monolithic Idea. But at the time—it was July of 1958—Kilby only hoped that his boss would let him build a model and give the new idea a try.

The boss was still an unknown quantity. It had been less than two months since Jack Kilby had arrived in Dallas to begin work at Texas Instruments, and the new employee still did not have a firm sense of where he stood. Kilby had been delighted and flattered when Willis Adcock, the famous silicon pioneer, had offered him a job in T.I.'s semiconductor research group, but the pleasure was tempered with some misgivings. Jack's wife, Barbara, and his two young daughters had been happy in Milwaukee. For that matter, Jack had prospered there. He had spent ten years in Milwaukee, at a small electronics firm called Centralab. During that decade, Kilby made twelve patentable inventions (including the reduced titanate capacitor and the steatite-packaged transistor) and one important discovery. He discovered the sheer joy of inventing. It was problem solving, really: you identified the problem, worked through 5 or 50 or 500 possible approaches, found ways to circumvent the limits that nature had built into materials and forces, and perfected the one solution that worked. It was an intense, creative process, and Jack fell in love with it. It was that infatuation with problem solving that had lured him, at the age of thirty-four, to take a chance on a new job in Dallas. Texas Instruments was putting him to work on the most important problem in electronics.

By the late 1950s the problem—the technical journals called it "the interconnections problem" or "the numbers barrier" or "the tyranny of numbers"—was a familiar one to the physicists and engineers who made up the electronics community. But it was still a secret to the rest of the world. At a time when the term "technological progress" had only positive connotations, Americans were looking ahead with happy anticipation to a near future when all the inventions of science fiction, from Dick Tracy's wrist radio to Buck Rogers's phone calls to Mars, would become facts of daily life. Already in 1958 a person could pull a transistor radio out of his pocket and hear news of a giant electronic computer that was decoding signals beamed at the speed of light from a miniaturized transmitter in a man-made satellite orbiting the earth at 18,000 miles per hour. Who could blame people for expecting new miracles tomorrow?

There was an enormous appetite for news about the future, an appetite that magazines and newspapers were happy to feed. Almost every month there was a report about some marvelous electronic innovation. First came the transistor, the invention that had given birth to this new electronic age—and then there was the tecnetron, the spacistor, the nuvistor, the thyristor. It hardly seemed remarkable when the venerable British journal *New Scientist* predicted the imminent development of a new device, the "neuristor," which would perform all the functions of the human neuron and so make possible the ultimate prosthetic device—the artificial brain. Late in 1956 a *Life* magazine reporter dug out a secret Pentagon plan for a new troop-carrying missile that could pick up a platoon at a base in the United States, "loop through outer space, and land the troops 500 miles behind enemy lines in less than 30 minutes." A computer in the missile's nose cone would assure the pinpoint accuracy required to make such flights possible. A computer in a nose cone? The computers of the 1950s were enormous contraptions that filled whole rooms—in some cases, whole buildings—and consumed the power of a locomotive. But that, too, would give way to progress. Sperry-Rand, the makers of UNIVAC, the computer that had leapt to overnight fame on November 4, 1952, when it predicted Eisenhower's electoral victory one hour after the polls closed, was said to be working on computers that would fit on a desk top. And that would be just the beginning. Soon enough there would be computers in a briefcase, computers in a wristwatch, computers on the head of a pin.

Jack Kilby and his colleagues in the electronics business—the people who were supposed to make these miracles come true—read the articles with a rueful sense of amusement. There were plans on paper to implement just about every fantasy the popular press reported; there were, indeed, preliminary blueprints that went far beyond the popular imagination. But it was all on paper, impossible to produce because of the limitation posed by the tyranny of numbers. The interconnections problem stood as an impassable barrier blocking all future progress in electronics.

And now, on a muggy summer's day in Dallas, Jack Kilby had an idea that might break down the barrier. Right from the start, he thought he might be on to something revolutionary, but he did his best to retain a professional caution; a lot of revolutionary ideas, after all, turn out to have fatal flaws. Day after day, working alone in the empty lab, he went over the idea, scratching pictures in his lab notebook, sketching circuits, planning a model; the more he studied the idea, the better it looked. When his colleagues came back from vacation, Jack showed his notebook to Willis Adcock. "He was enthused, but skeptical," Kilby wrote later. "I was very interested," Adcock recalled afterward. "But . . . what Jack was saying, it was pretty damn cumbersome; you would have had a terrible time trying to produce it." A test would require a model; that could cost $10,000, maybe more. There were other projects around, and Adcock was supposed to move ahead on them.

Jack Kilby is a gentle soul, easygoing and unhurried. A lanky, casual, down-home type with a big leathery face that wraps around an enormous smile, he talks slowly, slowly, in a quiet voice that has never lost the soft country twang of Great Bend, Kansas, where he grew up; that deliberate mode of speech reflects a careful, deliberate way of thinking. Adcock, in contrast, is a zesty sprite who talks a mile a minute and still can't keep up with his racing train of thought. That summer, though, it was Kilby who was pushing to race ahead. After all, if they didn't develop the idea, somebody else might hit on it. T.I. was hardly the only place in the world where people were trying to overcome the tyranny of numbers.

The Monolithic Idea occurred to Robert Noyce in the depth of winter—or at least in the mildly chilly season that passes for winter in the sunny valley south of San Francisco Bay that is known today, because of the idea, as "Silicon Valley." Unlike Kilby, Bob Noyce did not have to check with the boss when he got an idea; at thirty-one, Noyce was the boss.

It was January of 1959, and the valley was still largely an agricultural domain, with only a handful of electronics firms intruding on the peach and prune orchards. One of those pioneering firms, Fairchild Semiconductor, had been started late in 1957 by a group of physicists and engineers who guessed—correctly, as it turned out—that they could become fantastically rich by producing improved versions of transistors and other microelectronic devices. The group was long on technical talent and short on managerial skills, but one of the founders turned out to have both—Bob Noyce. A slender, square-jawed man who exudes the easy self-assurance of a jet pilot, Noyce has an unbounded curiosity that has led him to take up, at one time or another, hobbies ranging from singing madrigals to flying seaplanes. His doctorate was in physics, and his technical speciality was photolithography; at Fairchild, though, he became fascinated with the discipline of management, and he gravitated to the position of director of research and development. In that job Noyce spent a lot of time searching for profitable solutions to the problems facing the electronics industry; he thought about the optimum alloy to use for base and emitter contacts in double-diffuse transistors, about efficient ways to passivate junctions within a silicon wafer. And he also gave some thought, in the winter of 1958–59, to the tyranny of numbers.

Unlike the quiet Kilby, who does his best work alone, thinking carefully through a problem, Noyce is an outgoing, loquacious, impulsive inventor who needs somebody to listen to his ideas and point out the ones that couldn't possibly work. That winter Noyce's sounding board was his friend Gordon Moore, a thoughtful, cautious physical chemist who was another cofounder of Fairchild Semiconductor. "I spent a lot of time explaining to Gordon on the blackboard how you might do some of these things," Noyce recalled later.

Not suddenly, but gradually, in the first weeks of 1959, Noyce worked out the idea; on January 23, he remembers, "all the bits and pieces came together." He grabbed his lab notebook and wrote down The Monolithic Idea, in words quite similar to those Jack Kilby had entered in his notebook six months before: ". . . it would be desirable to make multiple devices on a single piece of silicon, in order to be able to make interconnections between devices as part of the manufacturing process, and thus reduce size, weight, etc. as well as cost per active element."

Like Kilby, Noyce felt fairly sure from the beginning that he was onto something important. "There was a tremendous motivation then to do something about the numbers barrier," he recalled later. "The [electronics] industry was in a situation—for example, in a computer with tens of thousands of components, tens of thousands of interconnections—where things were just impossibly expensive to make. And this looked like a way to deal with that. . . . I can remember telling Gordon one day that we might have here a solution to a real big problem."

At its core, the big problem that The Monolithic Idea was designed to solve was one of heightened expectations. It was hardly an unprecedented phenomenon in technologic history: a major breakthrough prompts a burst of optimistic predictions about the bright new world ahead, but then problems crop up that make that rosy future unobtainable—until a new breakthrough solves the new problem.

The breakthrough that gave rise to the problem known as the tyranny of numbers was a thunderbolt that hit the world of electronics at the end of 1947. It was a seminal event of postwar science, one of those rare developments that changes everything: the invention of the transistor.

Until the transistor came along, electronic devices, from the simplest AM radio to the most complex computer, were all built around vacuum tubes. Anybody old enough to have turned on a radio or television set before the 1960s may remember the radio tube: when you turned on the switch, you could look through the holes in the back of the set and see a bunch of little orange lights begin to glow—the filaments inside the vacuum tubes. A tube glowed because it was essentially the same thing as a light bulb: inside a vacuum sealed by a glass bulb, electric current flowed through a wire filament, heating the filament and giving off incandescent light. Experimenting with light bulbs at the beginning of this century, radio pioneers found that if they ran some extra wires into the bulb, it could perform two useful electronic functions. First, it could pull a weak radio signal from an antenna and strengthen, or amplify, it enough to drive a loudspeaker, converting an electronic signal into sound loud enough to hear; this made radio, and later television, workable. Second, a properly wired tube could switch rapidly—about 10,000 times per second—from on to off (because of its ability to turn a current on or off, the radio tube was known in England as a "valve"). This capability was essential for digital computers, which make logical decisions and carry out mathematical computations through various combinations of on and off signals.

But vacuum tubes were big, expensive, fragile, and power hungry. If a lot of tubes were closely connected in a single machine, as in a computer or a telephone switching center, all those glowing filaments gave off enormous quantities of heat that tended to transform expensive machinery into smoldering hunks of molten metal and glass—in effect, turning gold into lead. As we all know from the light bulb, vacuum tubes have an exasperating tendency to burn out at the wrong time. The University of Pennsylvania's ENIAC, the first important digital computer, never lived up to its potential because tubes kept burning out in the middle of its computations. The Army, which used ENIAC to compute artillery trajectories, finally stationed a platoon of soldiers manning grocery baskets filled with tubes at strategic points around the computer; this proved little help, because the engineers could never quite tell which of the machine's 18,000 vacuum tubes had burned out at any particular time. The warmth and soft light of the tubes also attracted moths, which would fly through ENIAC's innards and cause short circuits. Ever since, the process of fixing computer problems has been known as "debugging."

The transistor, invented two days before Christmas of 1947 by William Shockley, Walter Brattain, and John Bardeen of Bell Labs, promised to eliminate all the bugs of the vacuum tube in one fell swoop. The transistor was something completely new. It was based on the physics of "semiconductors"—elements like silicon and germanium that have unusual electronic characteristics. The transistor achieved amplification and rapid on-off switching by moving electronic charges along controlled paths inside a solid block (hence the term "solid state") of semiconductor material. There was no glass bulb, no vacuum, no warm-up time, no heat, nothing to burn out; the transistor was lighter, smaller, and faster—even the earliest models could switch from on to off about twenty times faster—than the tube it replaced.

To the electronics industry, it was a godsend. By the mid-1950s solid state was becoming the standard state for radios, hearing aids, and most other electronic devices. The burgeoning computer industry happily embraced the transistor, as did the military, which needed small, low-power, long-lasting parts for ballistic missiles and the nascent space program. The transistor captured the popular imagination in a way no other technological achievement of the postwar era had. Contemporary scientific advances in nuclear fission, rocketry, and genetics made awesome reading in the newspapers but were remote from daily life; the transistor, in contrast, was a breakthrough that ordinary people could use. The transistorized portable radio, introduced just in time for Christmas shopping in 1954, almost instantly became the most popular new product in retail history. It was partly synergy—pocket radios came out when a few pioneering disc jockeys were promoting a new music called "rock 'n' roll"—and partly sheer superiority: the first transistor radio, the Regency, was smaller, more power-efficient, far more reliable, and much cheaper ($49.95) than any radio had ever been before.

After indulging themselves for a year or two in the resentful skepticism with which academia generally greets revolutionary new concepts, physicists and electronics engineers gradually warmed to the expansive new possibilities offered by semiconductor electronics. By 1953 *The Engineering Index,* an annual compendium of scholarly monographs in technical fields, listed more than 500 papers on transistors and related semiconductor devices. A year later there were twice as many, a year after that even more. The titles reflected the intensity and the global scope of academic research spurred by the new technology:

"Interpretation of Alpha Values in p-n Junction Transistors"
"Le Transistron dans le circuit trigger"
"Circuito Multiplicatore del coefficient di risonanza con transistor"
"Tensoranalysis in Transistor-Rueckkopplungshaltungen"
"Perekhodnaya, chastotnaya, i fazovaya kharakteristika transistora"

But the papers reflected as well a quite unscholarly enthusiasm among the academics:

"Success Story—Transistor Reliability"
"Transistors Key to Electronic Simplicity"
"Méthodes d'Optimization Appliquées à la Microminiaturization"
"Fabulous Midget"

And then, as designers learned how to make use of the midget's fabulous properties, the tyranny of numbers began to emerge. Enthusiasm gave way to different emotions. By the late 1950s, when the problem had become acute, the titles in *The Engineering Index* reflected a general sense of disappointment, even despair, in the technical community:

"Switching Losses in Transistor Circuits"
Electronic Equipment—Weight and Volume Penalties to Flight Vehicles"
"Comment a été résolu le problème de la fabrication des transistors"
"Design Limitations of Semiconductor Components"

That last title was imprecise. The "design limitations" were not inherent in transistors; they stemmed from the basic structure of all electric circuits.

Building a circuit is like building a sentence. There are certain standard components—nouns, verbs, adjectives in a sentence; resistors, capacitors, diodes, and transistors in a circuit—each with its own function. A resistor is a nozzle that restricts the flow of electricity, giving the circuit designer precise control over the current at any point. The volume control dial on a TV set is really a resistance control. Adjusting the volume adjusts a resistor; the nozzle tightens, restricting the flow of current to the speaker and thus reducing the sound level. A capacitor is a sponge that absorbs electrical energy and releases it, gradually or all at once, as needed. A capacitor inside a camera soaks up power from a tiny battery and then dumps it out in a sudden burst forceful enough to fire the flashbulb. A diode is a dam that blocks current under some conditions and opens to let electricity flow when the conditions change. An electric eye is a beam of light focused on a diode; a burglar who steps into the light beam blacks out the diode, opening the dam to let current flow through the circuit to the alarm. A transistor is a faucet that can turn current flow on and off—and thus send digital signals pouring through the circuitry of a computer—or turn up the flow to amplify the sound coming from a radio's speaker. By connecting the standard components in different ways, one can get circuits, or sentences, that perform different functions.

Writers of sentences are taught to keep their designs short and simple. This rule does not apply in electronics. Some of the most useful circuits are big and complicated, with hundreds or thousands of components wired together. In the era of vacuum tubes, the designers' implicit awareness of power, heat, and size restraints set a limit to the scope of any circuit design; there was just no point in designing a machine that would melt to shards as soon as it was turned on. With the transistor, those fundamental design limitations disappeared. Now the designers could draw up plans for exotic communications and computer circuits using 50,000 or 500,000 transistors and similar numbers of diodes, resistors, and capacitors. On paper, these supercircuits could outperform anything that had been designed before. All you had to do was wire them together and—but that was the problem. That was where the numbers barrier came in. The new circuits were so big and complex it was virtually impossible to build them.

An electric circuit has to be a complete, unbroken path along which current can flow. That means that all the components of a circuit must be connected in a continuous loop: resistors wired to diodes, diodes to transistors, transistors to other resistors, and so on. Each component can have two, ten, even twenty interconnections with other parts of the circuit. Making the connections—wiring the parts together—was almost entirely hand labor: it was expensive, time-consuming, and inherently unreliable. A circuit with 100,000 components could easily require 1 million different soldered connections linking the components. The only machine that could make the connections was the human hand.

Even if somebody—the Pentagon, for example, where price, in the depths of the Cold War, was no object—could pay for that much hand labor, there was no way humans could put together a million of anything without turning out a few that were faulty. By the late 1950s the electronics industry had come head-to-head with this implacable limit. The Navy's newest aircraft carriers had 350,000 electronic components, requiring millions of hand-soldered connections; the labor cost—for wiring those connections and testing each one—was greater than the total cost of the components themselves. Production of the first "second generation" (i.e., completely transistorized) computer—the Control Data CD 1604, containing 25,000 transistors, 100,000 diodes, and hundreds of thousands of resistors and capacitors—lagged hopelessly behind schedule because of the sheer difficulty of connecting the parts. And new computers on the drawing boards would be far more complex. At the end of the decade, people were already planning the computers that would someday guide a rocket to a landing on the moon. But those plans called for circuits with 10 million components. Who could produce a circuit like that? How could it fit into a rocket?

"For some time now," wrote J. A. Morton, a vice president of Bell Labs, in an article celebrating the tenth anniversary of the transistor, "electronic man has known how 'in principle' to extend greatly his visual, tactile, and mental abilities through the digital transmission and processing of all kinds of information. However, all these functions suffer from what has been called 'the tyranny of numbers.' Such systems, because of their complex digital nature, require hundreds, thousands, and sometimes tens of thousands of electron devices." "Each element must be made, tested, packed, shipped, unpacked, retested, and interconnected one-at-a-time to produce a whole system," Morton wrote in a later article. "Each element and its connections must operate reliably if the system is to function as a whole. . . . The tyranny of large systems sets up a numbers barrier to future advances if we must rely on individual discrete components for producing large systems."

In essence, the small community of engineers exploring the frontiers of electronics in the 1950s faced the same abject frustration that had confronted the small community of seamen exploring the frontiers of navigation in the 1590s. At the far western extremity of the Atlantic, hard against the shores of Central America, the explorers could look westward from the masthead and see, "with a wild surmise," a vast new ocean, a whole new world, beckoning across the isthmus. But there was no way—no way short of the impossibly expensive, time-consuming, and unreliable voyage around the tip of South America—to get to that wonderfully promising new stretch of sea. The future was within sight, tempting, tantalizing, but out of reach. Just so for Jack Kilby, Bob Noyce, and their colleagues. A vast new electronic world was right there on the blueprints, but impossible to achieve. And so physicists and electronics engineers embarked on a great voyage of discovery, searching for a route across the numbers barrier.

The search became a top-priority technological concern throughout the industrialized world. The Royal Radar Establishment, racing to bring the honor of this important accomplishment to Great Britain, developed a promising concept as early as 1952 but failed to make it work. The French, the Germans, the Russians competed against one another; in the United States the Army, the Navy, and the Air Force competed just as fiercely, each service pushing its own preferred solution, each rejecting the ideas of the others. Private firms, sensing a gold mine, poured millions of dollars and man-hours into the effort. But through most of the 1950s none of these endeavors really helped. Patrick Haggerty, the president of Texas Instruments, complained that most of the proposed solutions to the tyranny of numbers "tend to exacerbate the tyranny."

The multifaceted efforts to deal with the numbers problem were grouped in the technical literature under the general title "miniaturization" (or "subminiaturization," or "microminiaturization"). It was an unfortunate term because it suggested a solution that could not work. The basic thrust of miniaturization was an effort to make electronic components extremely small, thus reducing the overall size and weight of complex electronic devices. This goal was obviously important to the military, which had to squeeze radios, radar and sonar devices, and computers into the nooks and crannies of missiles and submarines. One of the first miniaturization programs was a Navy-financed effort called "Operation Tinkertoy." But there were civilian implications as well. "In civilian equipment, such as computers," the trade journal *Electronics* noted, "the number of components alone makes miniaturization essential if the computer is to be housed in a reasonable-sized building."

But turning out transistors, resistors, and the like on Tinkertoy scale did nothing to reduce the sheer number of components and connections. The Tinkertoy business tended to exacerbate the tyranny because circuits composed of tiny parts were harder, and costlier, to build. On the assembly lines, the women who soldered circuits together—it was almost entirely women's work, because male hands were considered too big, too clumsy, and too expensive for such intricate and time-consuming tasks—now had to pick up miniature components and minute lengths of wire with tweezers and join them under a magnifying glass with a soldering tool the size of a toothpick. Circuits made under those conditions were far more likely to end up with faulty connections. In many cases, even a single bad connection could be fatal to the entire circuit, just as a single burnt-out bulb can make an entire string of Christmas lights go dark. Electronic devices that relied on circuitry employing the "microminiature" parts were famously unreliable.

To enhance reliability, the designers tried redundancy. Instead of building a radio with a single set of components (a typical small radio of the late fifties might have used a half-dozen transistors wired to a dozen resistors, capacitors, and diodes) the electronics companies started making radios with an extra circuit built right in—like a car built with two front axles just in case one should snap in half on the road. The redundancy business tended to exacerbate the tyranny because the extra components and extra wire required more interconnections, and thus more labor. Worse, redundant circuitry took up more space, and that was anathema, particularly to the people who built computers. Even without redundancy, electronic circuits were already too large. Large circuits undermined the single most important asset of modern electronic equipment: speed.

Calculators, computers, digital clocks, video games—for that matter, all digital electronic devices—are extremely dumb tools. But they are extremely fast extremely dumb tools. A computer reduces every question, every computation, every decision to the simplest possible terms: yes or no, one or zero, true or false (in the machine's internal circuitry, these two black-or-white states are represented by switches—transistors—that are either on or off). An astrophysicist mapping the universe in the observatory needs to calculate the twenty-fourth root of arctan 245.6; to do it, he types the problem into his computer. The machine has to work through two dozen separate yes-or-no steps—that is, transistors have to switch on and off two dozen times—just to figure out that someone has punched its keys. To determine which keys were pushed, and then to solve the problem, will take another 100,000 steps, quite possibly more. A kid playing Super Zaxxon in the arcade needs to destroy an enemy base; to do it, he pushes the "Fire" button. The machine has to work through two dozen separate yes-or-no steps just to figure out that the button was pushed. To fire the missile, and see if it hits anything, will take another 5,000 steps, quite possibly more. The machines can get away with their absurdly convoluted way of doing things only because the transistors switch, from on to off, from off to on, quickly. At a switching speed of once per second, computers would be impossible; at 1,000 times per second, merely impractical. Switching at a million times per second, computers become important. At a billion times per second—completing one step of the problem every nanosecond—they become the foundation of a revolution that has swept the world.

"After you become reconciled to the nanosecond," Robert Noyce has observed, "computer operations are conceptually fairly simple." In this respect, the electronic revolution of the twentieth century is the intellectual mirror image of the biological revolution of the nineteenth. Only after they became reconciled to enormously long periods of time—millions and millions of years, enough time for a dynohippus to turn into a donkey—could Darwin and his contemporaries contemplate species evolving on an evolving planet. Only after they became reconciled to enormously short periods of time—microseconds, nanoseconds, picoseconds—could the computer pioneers contemplate machines solving problems by turning switches on and off. The central concept of computer operations is that the machines operate inconceivably fast. Speed is the computer's secret weapon. If computers did not work as fast as they do, no one could justify the time and materials required to build them. At a switching speed of 1,000 times per second, it would take a whole second, maybe two, for a computer to add 2 and 2. At that rate, it would make no sense to buy the machine. The human brain, the sublimely intricate, powerful, efficient computer that everyone gets for free, can solve the problem faster than that.

The transistors in a computer switch on and off in response to electronic signals. A pulse of electricity moving through a wire reaches the transistor, and the transistor switches on; another pulse comes along, and the transistor switches off. No matter how quickly the transistor itself can switch, it cannot do so until the pulse arrives telling it what to do. The more wiring there is in a circuit, the farther these messenger pulses have to travel. In the 1950s, the limiting factor in computer speed was the travel time for those electronic signals moving through the circuit. In the biggest computers, with literally miles of wiring, it took so long for pulses to travel from one side of the circuit to the other that computation rates were seriously impaired.

At first blush, it might appear that there were two potential solutions to this problem: either speed up the signals, so they move through a large circuit faster, or shrink the circuits. Someday, if relativity theory is displaced and the laws of twentieth-century physics are stood on their heads, the first solution may be at hand. At present, however, it is against the law. Electronic pulses move through a circuit at the universal speed limit—the speed of light; if modern physics is correct, nothing will ever move faster. That left the second solution. To increase computing speed, it was necessary to reduce the distance the messenger pulses had to travel—that is, to make the circuits smaller. But smaller circuits meant decreased capacity. The result was a paradox. In the argot of the engineers, a computer's "power" is a measure of both its capacity to handle big problems and its speed in solving them. It was possible—at least in applications where cost and size were not serious problems—to increase problem-solving capacity by wiring in more transistors. But more transistors and more wire meant a larger circuit, which reduced computing speed. Thus the effort to build in increased computing power led to decreased computing power. It was the tyranny of numbers.

"It was a situation where, quite clearly, size dictated performance," Bob Noyce recalls. "Not just performance, in the sense of limiting computing speed, but the size and complexity of electronic circuits dictated cost, reliability, utility."

"The things that you could see that you wanted to do were going to take so many transistors, so many parts, that it would just be prohibitive, from a cost standpoint, from a size standpoint, any way you wanted to look at it," Jack Kilby remembers.

Noyce: "A large segment of the technical community was on the lookout for a solution."

Kilby: "There was just an awful lot going on. . . . It was pretty well accepted that this was the problem that had to be solved."

Noyce: "It was clear that a ready market awaited the successful inventor."

The successful invention—The Monolithic Idea—resolved the tyranny of numbers by reducing the numbers to one: a complete circuit would consist of one part—a single ("monolithic") block of semiconductor material containing all the components and all the interconnections of the most complex circuit designs. The tangible product of that idea, known to the engineers as the monolithic integrated circuit and to the world at large as the semiconductor chip, has changed the world as fundamentally as did the telephone, the light bulb, the horseless carriage. The integrated circuit is the heart of clocks, computers, cameras, and calculators, of Pac-Man and pacemakers, of deep-space probes and deep-sea sensors, of toasters, typewriters, and data transmission networks. The National Academy of Sciences has declared the integrated circuit the progenitor of "The Second Industrial Revolution." The first Industrial Revolution enhanced man's physical prowess and freed people from the drudgery of backbreaking manual labor; the revolution spawned by the chip enhances our intellectual prowess and frees people from the drudgery of mind-numbing computational labor. British physicist Sir Ieuan Madlock, formerly Her Majesty's chief science advisor, called the integrated circuit "the most remarkable technology ever to hit mankind." California businessman Jerry Sanders, founder of Advanced Micro Devices, Inc., offered a more pointed assessment: "Integrated circuits are the crude oil of the eighties."

All this came about because two young Americans came up with a new idea—or, more precisely, a not-so-new idea. In fact, the principle underlying the semiconductor revolution was one of the oldest ideas in electronics.

2

The Will to Think

In addition to the laws, rules, constants, principles, axioms, theories, and hypotheses they have devised to explain the mysteries of the natural world, scientists and engineers have developed a series of unwritten rules that purport to explain the mysteries of their business. Among the latter is a humorous, or perhaps quasi-humorous, principle sometimes referred to as "the law of the most famous." Briefly put, this natural law holds that whenever a group of investigators makes an important discovery, the most famous member of the group will get all the credit. This rule explains why the principle of thermionic emission came to be known as the Edison Effect.

Thermionic emission was observed for the first time in March 1883 in Thomas A. Edison's Menlo Park laboratory, when the inventor and his associates noticed something strange going on inside one of his first light bulbs. In addition to the electric current flowing through the carbon filament, there seemed to be another, separate current flowing through the vacuum inside the glass bulb—something quite impossible under contemporary explanations of electricity. Nobody at Menlo Park understood what was happening (the current was eventually found to be a flow of electrons boiling off the white-hot filament), but Edison wrote up the discovery and filed a patent for it. And then he set it aside. It was partly a matter of time and resources. Bedeviled by disputes with his creditors, legal battles over his patents, and frustrating efforts to improve his phonograph, microphone, and incandescent lamp, Edison was already working twenty hours a day, often more. But the main reason Edison abandoned the Edison Effect was that he saw no future in it. So what if current could flow through a vacuum? What good would that do?

Looking back a century later, when solid-state, or semiconductor, devices have proven superior to vacuum tubes in just about all electronic equipment, it can be said that Edison had a point. Hindsight suggests that the fifty-year period of electronics development based on the vacuum tube was basically a digression. By 1883, electrical pioneers such as Edmond Becquerel, Ferdinand Braun, and Michael Faraday had found that certain substances—known today as semiconductors—had a variety of useful electronic characteristics. Had the early work on these materials been continued, it is not too great a flight of fancy to suggest that the modern semiconductor revolution might have come a half-century sooner. But after the discovery of the Edison Effect, electronics research took a new direction—with dramatic results. The work at Menlo Park led, fourteen years later, to the experiment known as "the zero hour of modern physics"—the discovery of the electron—and from there, along a more or less straight line, to wireless telegraphy, radio, television, and the first generation of digital computers.

In 1883, however, all that was considerably less than obvious, even to an exceptional visionary like Edison. At first, the Edison Effect held interest only for scientists—and scientific value was not a commodity of great importance to Thomas A. Edison. "Well, I'm not a scientist," the Wizard of Menlo Park said. "I measure everything I do by the size of the silver dollar. If it don't come up to that standard then I know it's no good."

That line was classic Edison. All his life he portrayed himself as the supreme pragmatist, relying on hard work and common sense to build a better life for his fellow man—and get rich in the process. It was an archetypal American picture, and essentially an accurate one, for Edison's life has the elements of the classic American story. The seventh child of an infrequently successful businessman, he grew up in medium-size towns in Ohio and Michigan. He had about four years of school, counting the time his mother tutored him at home, and set out at the age of twelve to make his fortune. He sold snacks on the Detroit–

Port Huron train. He started a newspaper called *Paul Pry*. He fell into and out of numerous jobs as a telegraph operator, a profession he enjoyed so much that he called his first two children "Dot" and "Dash." Tinkering continually with his employers' telegraphic equipment, he began to design useful improvements and gradually discovered that he could make a living as an inventor. By his thirty-fifth birthday he was a millionaire, a leader of industry, and probably the best-known man on Earth. When he announced early in 1878 that he was thinking about an electric light, illuminating gas stocks plummeted on Wall Street. When the *New York Daily Graphic* reported that Edison had invented a machine that spun food and wine from mud and water, many newspapers failed to notice the April 1 dateline and ran the story straight—just one more miracle from Menlo Park. When Edison died, at eighty-four, in 1931, someone proposed that all lights be turned out for two minutes as a memorial. The idea was dropped on the ground that it would be impossible for the world to function that long without the electric light.

Despite fame and fortune, Edison remained an uncouth hayseed who flaunted his disdain for cleanliness, fashion, order, religion, and science. A journalist touring the famous Menlo Park laboratory in 1878 described the proprietor this way: "The hair, beginning to be touched with gray, falls over the forehead in a mop. The hands are stained with acid, his clothing is 'ready-made.' He has the air of a mechanic, or with his particular pallor, of a night-printer." Edison worked in 40-, 80-, or 100-hour bursts, catching a few intervals of sleep under a bench in the lab. He adopted the motto *Perseverentia omnia vincit,* a phrase that he subsequently translated to "Genius is one percent inspiration and 99 percent perspiration." He had absolute confidence in this formula; he was certain he could solve any problem if he just tried enough solutions. After an assistant noted wearily that the Menlo Park team had worked through 8,000 different formulas in a futile effort to build a storage battery, the inventor replied that "at least we know 8,000 things that don't work." Struggling to find an efficient filament for his incandescent light, Edison decided to try everything on earth until something worked. He made a filament from dried grass, but that went haywire. He tried aluminum, platinum, tungsten, tree bark, cat's gut, horse's hoof, man's beard, and some 6,000 vegetable growths before finding a solution in a carbonized length of cotton thread.

Publicly, at least, the great empiricist had no truck for scientists, mathematicians, or any college graduate—"filled up with Latin, philosophy, and all that ninny stuff." With great fanfare, he invented an "ignorameter" to prove that intellectuals had no common sense. When he hired a "science advisor" of his own—Francis Upton, holder of a Ph.D. from Princeton and student of the great German physicist Helmholtz—Edison promptly named him "Culture," a term of derision among the gang at Menlo Park. Soon enough the newspapers were reporting how Edison had asked his man Culture to determine the volume, in cubic centimeters, of an empty glass bulb. Upton set to the task with a complex series of differential equations that would lead, after a few hours, to a close approximation of the correct figure. With a smirk, the pragmatic inventor grabbed the bulb, filled it with water, and poured the water into a measuring cup—determining the precise volume in a matter of minutes.

Culture eventually became acculturated to life in the lab, and Edison eventually came to realize that Upton's mathematical and physical skills were an important asset. Upton designed and built the dynamo—in modern terms, the generator—that provided safe, efficient power for the Edison lighting system and the Edison electric train. Upton's skills as an observer and theoretician were so central to the discovery of thermionic emission that the phenomenon might more fairly be called the "Upton Effect."

When the electric light was still in its birth throes, Edison and Upton noticed that, over time, the clear glass bulb tended to grow black. Edison and some assistants set out on a characteristic series of trial-and-error experiments to eliminate the problem. Upton, meanwhile, made careful observations of the phenomenon. In March of 1883, he suggested putting a small metal plate inside the bulb. This didn't help (the problem of darkening bulbs was eventually solved by using purer materials for the filament), but it led to a fascinating discovery. Edison and Upton found, to their surprise, that current was flowing in the metal plate. Where in the world could this current come from? Current was flowing through the filament, of course, but there was no connection whatsoever between the filament and the plate. On a hunch, the researchers increased the current through the filament—and found that the current in the metal plate increased proportionally. Evidently electric current was flowing *across the vacuum* from filament to plate. This was quite astonishing—at least to Upton, who trusted scientists—because the experts had established that electric current could not traverse a vacuum.

Over the next few years, the mysterious current-in-a-vacuum—it was soon being called the "Edison Effect"—became a prize exhibit at electrical exhibitions on both sides of the Atlantic. John A. Fleming, a scientist on the staff of Edison's British subsidiary, ordered some Edison Effect lamps and tried a number of experiments. Edison had always used a direct current of electricity in his work—a current in which the electricity flowed in the same direction all the time. During the winter of 1884–85, Fleming tried something different. He hooked up the filament to a generator that produced alternating current—a current that constantly changes direction, back and forth, back and forth, as often as 120 times per second. He was mystified to see that, even with an alternating current racing back and forth through the filament, the current flowing to the metal plate was still direct current, never changing direction. The lamp had converted alternating current to direct current.

No one could explain this result—least of all Edison, of course, who stood aside and smiled as the scientists did their stuff. Research on the Edison Effect was merely aesthetics, Edison wrote to a friend, and "I have never had time to go into the aesthetic part of my work. . . .

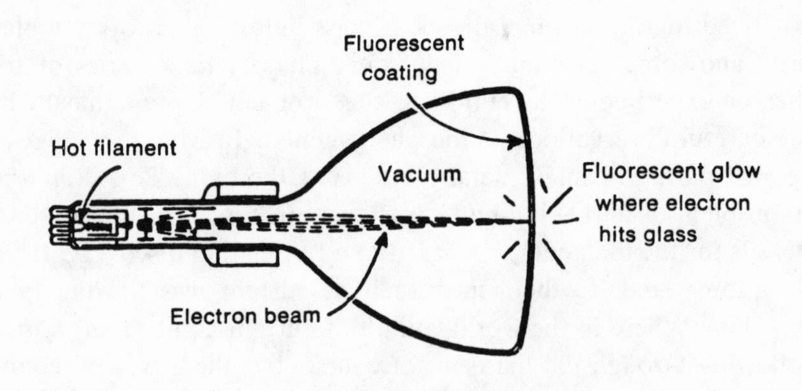

How a cathode ray tube works.

But it has, I am told, a very important bearing on some laws now being formulated by the Bulged-headed fraternity of the Savanic world."

It did indeed. Savants in the United States and Europe undertook extensive experimentation on the flow of electricity through a vacuum. The basic apparatus for this work was an elongated glass tube with a piece of carbon or metal at one end that was heated—just like the filament in Edison's light bulb—until the thermal energy emitted an electric current (hence "thermionic emission") through the vacuum. The piece of metal that emitted the current was called a "cathode," so the glass tube was known as a "cathode ray tube." The current would beam down the tube to the far end; at the spot where the beam hit, the glass would phosphoresce, or glow. At the time, this exotic apparatus was found only in the finest laboratories in the U.S. and Europe. Today it is found in living rooms, basements, and bars everywhere, in the form of the television picture tube.

The savants hoped that the tube would provide a clear enough picture of cathode rays to permit an explanation of the electric force. The mystery of electricity had prompted a number of contradictory hypotheses. Early researchers had postulated that electricity was a fluid —which is why we still talk today of "current" and "flow." By the 1880s this notion had given way to a pair of competing theories. One view held that electricity was a wave phenomenon, like sound and light; the other school of thought considered the electric beam in the cathode ray tube to be a stream of separate particles. Wave or particle? The greatest minds in physics pondered, debated, speculated over the question. The answer finally came from one of the most fascinating and formidable intellects in the history of physics—Prof. Sir Joseph John Thomson.

J. J. Thomson was born in 1856, the son of a bookseller in Manchester. The family's plan had been that J.J. would be apprenticed to a local engineer and take up that profession. But his father's death, when J.J. was sixteen, left the Thomsons unable to pay the fees engineers charged for training apprentices. The boy won a scholarship at a local college and quickly came under the spell of mathematics and physics. The timing was perfect—Thomson's professional career spanned the most fertile era in physics since Newton's day—but nobody knew that during his college days. Quite the contrary, in fact: in the 1870s, there was a general sense that the interesting part of physics was over, and all that remained was refining the measurements. After all, everyone knew that all matter was made of indivisible particles called atoms, that differences among elements were due to differences in their atoms. The great theoretician James Clerk Maxwell, on the august occasion of his nomination as Cavendish Professor at Cambridge—in essence, the physicist laureate of England—noted the consensus "that in a few years all the great physical constraints will have been approximately estimated, and that the only occupation which will then be left to men of science will be to carry on those measurements to another place of decimals." Already scientists had measured the mass of the smallest object in the universe—the hydrogen atom, weighing about .0000000000000000000000017 grams.

Thomson went to Cambridge in 1876, studied under the great Maxwell, and quickly distinguished himself. He won the Trinity Prize for the interesting discovery that a drop of water would not evaporate if given an electric charge. In an 1881 paper he suggested that there was a connection between the mass and the energy of a moving sphere —an idea that was crystallized by another physicist twenty-four years later in the equation $E=MC^2$. As the preeminent Cambridge physicist, Thomson succeeded in 1884 to the chair—both the professorship and the piece of furniture—that Maxwell had held. He was twenty-eight years old. "Things have come to a pretty pass," an older colleague groused, "when mere boys are appointed professors."

"J.J. spent a good part of most days sitting in the armchair that had belonged to Maxwell, doing mathematics," a student recalled. He sat there thinking, working out problems, because he had absolute confidence in the power of thought. He was certain that the human mind, aided by mathematics, could comprehend all physical phenomena. But the professor's wide-ranging curiosity reached far beyond math and physics. His workday ran, at the most, from ten in the morning to six at night (with a break for afternoon tea), and he regularly found time to cheer for Cambridge at crew and rugby matches. He read Housman's poetry in manuscript and made it a point never to miss opening night of a new Gilbert and Sullivan operetta. He was fascinated by gardening, golf, and American politics. He was British to the core. In his memoirs he notes with great pride that twenty-seven of his students (including his son) were elected to the Royal Academy; as an aside, he mentions that seven of them (including his son) also picked up Nobel Prizes. J.J.'s own Nobel Prize, in 1906, seems to have satisfied him less than the knighthood he received two years later. When he died, at eighty-four, in 1940, he was buried in Westminster Abbey near the grave of Isaac Newton.

Physicists talk about Thomson's cathode ray experiment of 1897 the way architects talk about Dulles Airport. The approach was elegant, the conclusion bold and stunning, and the result left an indelible mark on everything that came after. In essence Thomson decided he could ascertain the nature of the electric beam shooting through the cathode ray tube by subjecting it to different forces and measuring the results as precisely as possible. His experiments were quite beautiful to watch. By placing magnets around the glass tube, Thomson could make the cathode ray spin in a perfect spiral. Since the vacuum in the tube was not complete, the cathode ray would occasionally strike a stray gas atom and give it an electric charge. The charged atom, or "ion," would glow with a colorful splendor until it was neutralized, or "recombined," by another atom. So lovely were the results that Thomson and his colleagues wrote a song about it all, to the tune of "Clementine":

In the dusty lab'ratory,
'Mid the coils and wax and twine,
There the atoms in their glory
Ionize and recombine.

Chorus: Oh my darlings! Oh my darlings!
Oh my darling ions mine!
You are lost and gone forever
When just once you recombine.

In the weird magnetic circuit
See how lovingly they twine,
As each ion describes a spiral
Round its own magnetic line.

Chorus

In a tube quite electrodeless,
They discharge around a line,
And the glow they leave behind them
Is quite corking for a time.

Chorus

At first, Thomson's research produced nothing but confusion; his findings did not seem to fit any single theory. The fact that the electric beam could be bent by a magnet or an electric field strongly suggested that it consisted of particles—for no wave was susceptible to magnetic attraction. But there was also compelling evidence against the particle hypothesis. There was no indication, even with the most exacting measurements, that the beam was deflected by gravity, as any stream of particles should be. Further, the beam could pass through a thin sheet of metal foil without leaving a hole. It was known that certain energy waves did that—just as light waves pass through a window without leaving a mark on the glass—but it seemed impossible for a solid particle to do so.

To resolve these contradictory results, Thomson fell back on mathematics. Through a series of ingenious calculations, he determined the velocity of the moving beam—it traveled about 20,000 miles per second, infinitely faster than any object had ever been found to move before. Using that result, he calculated what the mass of the cathode ray particles would be, if indeed they were particles. This produced a result that was simply impossible. The mass of each particle would be less than one one-thousandth the size of a hydrogen atom. As a physical matter, this could not be right, for the hydrogen atom was well known to be the smallest of all material objects, and quite indivisible. As a mathematical matter, though, it could not be wrong. At this point, Thomson ended his experiments. There was nothing left to do but sit down in Maxwell's armchair and think the matter through.

The conclusion J.J. drew from this research is so elementary today that it is almost impossible to appreciate the enormous intuitive leap, the sheer revolutionary daring, that was required to state it in 1897. In one fell swoop, Thomson split the unsplittable atom and explained the inexplicable force of electricity. He presented his wholly new picture of the physical world in an informal speech to The Royal Institution on a Friday evening in April.

"I have lately made some experiments which are interesting," he said. The atom was not indivisible. He had found within the atom a new kind of particle—a particle at least a thousand times smaller than any atom. These subatomic particles—Thomson called them "corpuscles," but eventually the term "electrons" was adopted instead—are universal constituents of all matter, found in every atom. The electrons are magnetic and carry a negative charge, which explains why a magnet or an electric field made them swerve from their straight path. Their mass is so minute that the force of gravity upon them was undetectable. They are so much smaller than any atom that they could slip through the open spaces in an atom of metal—and thus shoot through a sheet of metal foil without a trace. The cathode ray—for that matter, any electric current—consists of a stream of these charged particles.

In subsequent lectures and papers, Thomson and his colleagues refined their picture of the electron and postulated other subatomic particles. Since the electrons carried a negative charge, there must be another, positively charged particle in the atom; equal numbers of electrons and protons (as the positive particle came to be called) would make the ordinary atom electrically neutral. Because of its small mass, the electron can move about. Indeed, if a piece of metal is given an electric charge or heated to incandescence (as in the Edison Effect), electrons will stream away in huge numbers. This stream of moving electrons, Thomson concluded, is an electric current.

Although Thomson's leap of insight developed from experiments on electric current in a vacuum, his explanation of electric current—a flow of moving charges—is equally valid for current in a block of semiconductor material. If scientists had gone back to the semiconductor work that had largely been laid aside a generation earlier, the next important developments might have come in that field. As it happened, though, J.J.'s discovery was first applied in vacuum tubes—and the tube became the central component of electronic devices for the next half-century. This came about because of the work of John Ambrose Fleming and Lee De Forest.

J. A. Fleming was a contemporary of Thomson's. Like Thomson, he studied physics at college only because his family could not pay for an engineering apprenticeship. Like Thomson, Fleming was a distinguished professor (at University College, London) and a member of the leading scientific societies. Unlike Thomson, Fleming was interested in making money from his work; one result was that, like Edison, he became ensnarled in endless patent litigation. In pursuit of an income, Fleming worked the lecture circuit, traveling all over the British Isles to give scientific demonstrations.

Hard-working, highly disciplined, extremely demanding of himself and those around him, Fleming was determined that everything about his lectures should be perfect—he rehearsed with a stopwatch so that every word and gesture would come at the right second—and thus failed to see the humor in a prank perpetrated during a widely advertised talk he gave in 1903. To demonstrate the wonders of wireless telegraphy, Fleming had arranged to receive a long-distance message in mid-lecture from the great Marconi himself. At the appropriate time, the telegraph key began to rattle—but instead of the weighty words Fleming had chosen to mark the historic occasion, the message was an off-color limerick ("There was a young fellow of Italy/Who diddled the public quite prettily . . ."). It turned out that a playful student in the audience had brought a transmitter of his own for purposes of mischief. Fleming, thoroughly unamused, wrote a thundering letter to *The Times* to denounce such "scientific hooliganism."

Fleming's interest in money also led him to pioneer a practice common among modern faculty members—consulting to industrial concerns. In 1882 he was appointed "electrician" (the modern term would be "science advisor") to the Edison Electric Light Company, Ltd. He held that position only a few years, but they happened to be the years when Edison and Upton were working on the Edison Effect. Fleming, as we have seen, made the interesting discovery in 1884 that the Edison Effect current—the current from the filament to the metal plate—never changed direction, even when alternating current was sent through the filament. Years later, after Thomson had established that the current is a flow of electrons, Fleming was able to explain why. Electrons boiling off the hot filament flowed to the metal plate. But the plate was not hot enough to emit electrons, so no current flowed back from plate to filament. Thus the Edison Effect always produced direct current.

At the turn of the century Fleming landed another consulting position, this time with Marconi's Wireless Telegraph Company, Ltd. Radio in this primordial era was still as much a toy as a tool, and there were several problems facing the Marconi firm. One was the inability to tune the radios to a specific frequency, an improvement that could have prevented pranksters from sending limericks in place of important messages. A more significant obstacle to serious radio transmission was the absence of a reliable rectifier.

A radio transmitter beams out signals that travel through the sky in the form of alternating current. But the receiving instruments that turn those signals into information—a telegraph key, for example, or a radio's speaker—operate on direct current. The crucial need, then—the missing link—was a device that could take the alternating current sent by the transmitter and convert, or "rectify," it to a direct current that echoed the pulsations in the original signal.

The materials known today as semiconductors have this rectifying quality, and the first radios employed a semiconductor crystal to rectify the signal. Such radios came to be known as "crystal sets." Since nobody knew much about semiconductor technology, crystal sets seemed to work only when they chose to, and were too capricious for business use. If radio was to have any serious impact on the world, someone would have to find a dependable way to convert the alternating radio signal into direct current. This was the task Marconi assigned to J. A. Fleming.

Fleming initially tried to get reliable rectifying action out of the standard crystal. In October of 1904, however, he realized that this approach would not be fruitful and began thinking hard about other devices that could convert an alternating current to direct current. What mechanism was there that always produced direct current? Suddenly, he recalled his experiments of twenty years before. Fleming himself described the moment of discovery:

> I was pondering on the difficulties of the problem when my thoughts recurred to my experiments in connection with the Edison Effect.
>
> "Why not try the lamps?" I thought.
>
> . . . I went to a cabinet and brought out some lamps I had used in my previous investigations. . . . I started the oscillations in the primary circuit. To my delight I saw the needle of the galvanometer indicate a steady direct current. . . . We had in this particular kind of electric lamp a solution to the problem of rectifying high frequency wireless currents. The missing link in wireless was found—and it was an electric lamp.

In addition to its usefulness for radio, Fleming noted that the current flowing from filament to plate could switch off and on far more rapidly than any mechanical switch. "So nimble are these little electrons," Fleming noted, "that however rapidly we change the electrification . . . the plate current is correspondingly altered, even at the rate of a million times per second." Because of this capacity to turn current on and off, Fleming called the modified light bulb a "valve." In the technical literature, Fleming's rectifying lamp is called a "diode," because it has two electrodes—the filament and the plate.

The Fleming diode made possible the production of dependable radio receivers, but there was still another problem to be resolved before radio became a practical instrument for sending information over any appreciable distance. Radio beams attenuate as they travel; the farther a signal has to go, the weaker it gets. After a signal had traveled 30 miles or so it was too weak to drive any kind of microphone; a slightly longer distance so diminished the current that it could barely move a telegraph key. What was needed was a device in the receiver that could strengthen, or "amplify," the incoming signal without distorting its pattern of pulsation and modulation. The need was met, two years after Fleming's invention, in the cluttered New York office of Lee De Forest.

De Forest was the son of a Congregationalist minister who moved, shortly after the Civil War, from the Midwest to Talladega, Alabama, to run the Negro college there. Stuck with this double whammy—a Yankee who lived with the Negroes—young Lee had few friends among his fellow whites in Talladega and spent his childhood reading science. He did indeed become a scientist; he took a Ph.D. at Yale after writing what was probably the first American dissertation on radio waves. Still, there was considerably more of Edison in him than of Thomson. A tireless self-promoter, De Forest worked feverishly all his life to attain the wealth and fame he felt he deserved. It was often an uphill battle. De Forest spent huge sums, with indifferent results, in legal battles over patents. He spent two years at the height of his career, 1912–13, fighting a federal mail fraud indictment resulting from his prediction, in a letter to potential investors, that the human voice would someday be broadcast across the Atlantic. No one could possibly believe such "absurd and deliberately misleading statements," the prosecutor declared. The jury did, and De Forest was acquitted.

Not a shrinking violet, De Forest yielded to none in his esteem for his scientific accomplishments. Asked to discuss his radio amplifier before the Franklin Institute, the inventor assured the assembled engineers that "a more revolutionary step was never taken in the history of engineering." De Forest titled his autobiography *The Father of Radio,* an immodesty that prompted Isaac Asimov to note that "few inventions have had so many fathers." Just before he underwent delicate cancer surgery at the age of eighty-five, De Forest overheard the doctors saying the tumor would be removed by electrodesiccation. "Commonly known as the hot wire," De Forest croaked from the operating table. "I invented it in 1907."

De Forest's most important invention, the radio amplifier, was based on a fundamental principle of electricity: unlike charges attract, and like charges repel. An object carrying a negative charge is attracted toward a positive charge just as a paper clip is pulled toward a magnet. In a rainstorm, when clouds and earth develop opposite charges, the attraction is strong enough to pull a lightning bolt of electrons across the gap. This principle also explains the paparazzi's favorite phenomenon, static cling. As the starlet scoots over to get out of the limousine, the hem of her dress rubs some electrons off her nylon stockings. With these excess electrons, the dress acquires a negative charge; the stockings become positive. Negative hem clings to positive stocking—at about mid-thigh, if the photographers are lucky—and a thousand flashbulbs pop as she leaves the car.

Tinkering with some early rectifying tubes in 1906, De Forest put a wire screen between the filament and the metal plate. Normally, this did not affect the Edison Effect current because electrons flowed right through the open screen to the plate. But when De Forest put a negative charge on the wire screen, like charges repelled: the negative screen repelled the negatively charged electrons, and current flowing to the plate was sharply reduced. When he sent a positive charge to the screen, unlike charges attracted: the screen attracted electrons, and the current to the plate was increased.

Experiments showed that a small change in the charge on the metal screen caused a big change in the current flowing to the metal plate. More important, the variations in the Edison Effect current exactly mimicked variations in the current sent to the wire screen. Here, then, was a precise amplifying mechanism. If the weak current from a distant radio signal was sent to the wire screen, it shaped a much stronger current that precisely matched the fluctuations of the radio beam. This stronger current could drive a telegraph key or loudspeaker. De Forest called the metal screen a "grid" and filed a patent entitled "Device for Amplifying Feeble Electrical Currents."

Since the De Forest tube had three electrodes—the filament, the metal plate, and the grid—it was technically known as a "triode." The triode amplifier made radio a practical everyday reality. With further advances—radio companies eventually developed tetrode and pentode tubes—performance was greatly enhanced. By 1930 radio had swept the world (De Forest's absurd prediction of trans-Atlantic broadcasting had come true in 1915). By the mid-forties engineers had learned how to use radio signals to draw a picture on a glass tube; the result was a device known at first as an "iconoscope" but today as television. Computer pioneers took advantage of the tube's rapid switching in building the first generation of digital computers.

With the capacity to perform three essential functions—rectification, amplification, and rapid switching—the vacuum tube, at heart just a souped-up light bulb, was the hub of a new electronic world. If the development of electronics were viewed as a battle of competing technologies, vacuum tubes had overcome semiconductor devices, like the crystal set, and left them far behind. But this battle was not yet over.

The renaissance of the semiconductor began in the late 1930s, spurred by growing fear on both sides of the English Channel that war was imminent. Recognizing that the coming conflict would depend largely on air power, scientists in England and Germany raced to develop an early warning system, using radio techniques, to spot approaching enemy planes. Fortunately for the allies, the British perfected the concept first. Since their system could not only find planes but also gauge their distance, or range, from England, the British called the invention "Radio Detection and Ranging." This name was quickly shortened to the acronym "radar."

A radar station shot a radio beam into the air. As long as it didn't run into anything, the beam kept moving in a straight line at the speed of light (some radar beams sent off during the Battle of Britain are presumably still moving out through space today, about 43 light-years from Earth). But if the signal hit a piece of metal—say, a Luftwaffe bomber—in midair, the beam would bounce back to the radar station like a tennis ball bouncing off a wall. By marking where the returning beam came from, and measuring how long its round trip had taken, the British defenders could tell their fighters where to intercept the enemy.

At first the British had hoped to use standard radio equipment to transmit and receive the radar beams. This didn't work, because vacuum tube rectifiers could not handle the high-frequency signals required for radar. Desperate for some other rectifying apparatus, the engineers went backward in history and resurrected the crystal set. Crystal rectifiers had been around for decades, but they had never been significant because their performance was always iffy. By the late 1930s, however, much more was known about the crystals in these crystal sets—the elements known as "semiconductors." The radar engineers, working from this new base of knowledge, were able to build reliable crystal receivers.

The deeper understanding of semiconductors had come about in a random, almost haphazard, manner. Unlike the development of vacuum tube technology, in which each new researcher and each new discovery seemed to lead neatly ahead to the next, the scientific world's knowledge of semiconductors grew out of a disjointed series of experiments and hypotheses. A monograph published in Berlin, an interesting experiment in Cambridge, a suggestion from Paris, a countertheory from Copenhagen—all this work gradually came into focus in the 1930s. Eventually, two lines of scientific work, one experimental and one theoretical, merged into a single theory of semiconductor physics.

The experimental contribution sprang from commercial roots. As electrical power became a commercially important commodity, the firms that sold and used electricity had to know the most efficient way to transmit it. If a company had to build a power line, should the line be made of copper or cotton? To answer that, experiments were run to measure conductivity of countless different materials. Conductivity—that is, how easy it is for an electric current to flow through a given material—is a physicist's concept; it is the opposite of the electrical engineers' concept of resistance. The electricians' unit of resistance is called the "ohm." Some physicist with a wry humor accordingly decided that the unit of conductivity should be called the "mho."

In experiments to determine mho ratings for hundreds of materials, certain patterns emerged. In some substances, particularly metals like silver, gold, and copper, current could flow easily. These materials were labeled "conductors." Other substances—quartz, glass, rubber, and wood are prominent examples—blocked current flow. They were called "insulators." Between these two extremes was a class of materials that conduct better than insulators but not as well as conductors. These semi-good conductors—elements like selenium, germanium, and silicon —were given the generic label "semiconductors."

Still lacking, though, was an explanation of why some materials made better conductors than others. Why did electrons travel so readily through copper but so reluctantly through glass? And what was it about silicon that made it fall in between? The answer to that question was provided by the theorists of quantum mechanics—and particularly by a quiet, deferential Dane, Niels Henrik David Bohr, who worked out the basic architecture of the atom.

Niels Bohr was one of the great people of the twentieth century—not only one of the most powerful intellects but also one of the most generous and humane of men: "the incarnation of altruism," his friend C. P. Snow wrote. Born in Copenhagen in 1885, the son and grandson of distinguished academicians, he grew up in highly intellectual surroundings. The family read widely in four languages and was steeped in music and the arts. Niels was also a soccer ace, although he was not picked for the Danish Olympic team (his brother snared a spot in 1908). Denmark did come beckoning a few years later, however, when Bohr had emerged in the top ranks of physicists and was working in England. To lure Bohr home, the Danes established an Institute of Theoretical Physics in Copenhagen (the Carlsberg brewery put up the money). Under Bohr's direction the institute became, for a while, the world capital of quantum physics.

"He was not, as Einstein was, impersonally kind to the human race," C. P. Snow wrote after Bohr's death in 1962. "He was simply and genuinely kind. It sounds insipid, but in addition to wisdom he had much sweetness." His selfless nature shone through radiantly during World War II. He donated a priceless possession—his gold Nobel Prize medal—to be melted down so the proceeds could go to Finnish war relief. At considerable professional and personal danger, he spoke out forthrightly against the Nazis and worked secretly to help Jewish scientists escape the Third Reich. Finally forced to flee his homeland in a small fishing boat with his son Aage (a chip off the block who was to win a Nobel Prize of his own), Bohr then set up an underground railway to spirit Jews out of Denmark. After the war, he began a tireless campaign against further deployment of nuclear weapons.

Midway through his scientific career, Bohr made a fascinating intellectual conversion. As a young man, he was convinced that the world around him had a logical order, that natural phenomena could be explained with rigorous logic and reason. He fell away from the church because of his feeling that its doctrines were logically untenable. Later, though, as he discovered parts of the world where logic evidently did not govern, he accepted a somewhat muddier picture. By the mid-1920s he had developed a "Principle of Complementarity," which held that physics was large enough to contain some seeming illogic and contradictions. Late in life he designed a personal coat of arms that carried the yin/yang symbol and the motto *Contraria sunt complementa.*

Bohr took his Ph.D. at Copenhagen in 1911, writing his dissertation on the still new concept of the electron, and then headed off to Cambridge to study under J. J. Thomson himself. There he became familiar with Thomson's conception of the structure of the atom—a theory known as the "raisin cake atom" because it posited a sort of sponge cake with electrons scattered about here and there like raisins. Next Bohr went to Manchester to work with another great physicist, Ernest Rutherford; there he learned of Rutherford's "nuclear atom," which posited a small nucleus set inside an amorphous cloud of electrons. Then, in 1913, Bohr set down his own picture of the atom—a hypothesis that has prevailed, with regular refinements, ever since.

The "Bohr atom" is the atom that most adults today saw in their high school science books: the "solar system" model, with electrons swirling in concentric orbits around a central nucleus. "In this picture," Bohr explained in his Nobel Prize address in 1922, "we see a striking resemblance to a planetary system, such as we have in our solar system." The key point of the Bohr picture, though, was his insistence that electrons could not orbit in just any old spot. Using quantum mechanics and some mind-boggling mathematics, Bohr determined precisely how far from the nucleus each orbit should be, and how many electrons can reside in each orbit.

Under these rules, the orbit that is farthest from the nucleus can have from one to eight electrons. The electrical characteristics of each element are determined by this outermost orbit—specifically, by the number of electrons in the outer orbit. If an atom has only one electron in the farthest orbit, that electron will not be tightly bound to the nucleus; it could easily break away. But in an atom with a full house—eight electrons—in the outer orbit, the electrons will be held tightly in place.

Working from this picture of the atom, quantum physicists could predict from their theory which materials would be good conductors of electric current. A substance that easily releases electrons would supply the free-flowing electrons that make up electric current; such a substance should be a good conductor. In a material that did not release free electrons, current would not flow; it would be an insulator. Theoretically, then, the conductivity of any material would be determined by the number of electrons in its outermost ring.

No one has ever seen an atom. Until we do, parts of the quantum picture of atomic structure will remain, in a strict sense, merely theory. The quantum view of conductivity, however, can be tested, because of the experiments that determined the conductivity of specific materials. When these experimental results are compared with the predictions of quantum theory, theory and experiment match perfectly.

The materials found to be the best conductors—silver, copper, gold—are indeed elements with a single electron in the outermost orbit. Materials that have proven the best insulators are indeed those with eight outer electrons. As a general matter, elements with three or fewer outer-ring electrons are conductors, and those with five or more are insulators. At the precise center of this continuum stand the semiconductors. Semiconductors, such as silicon and germanium, have four electrons in the outermost ring.

It is this special feature of semiconductor materials that makes them so spectacularly useful in electronics. Sitting on the fence, perched midway between the conductors and the insulators, semiconductors can perform valuable electronic service precisely because of their in-between structure.

Because the semiconductors are right on the border line between conductors and insulators, men have found ways to alter their conductivity. This is done by a process called "doping." Here's how it works: In a solid block of pure silicon, the individual atoms tend to link up with their neighbors. Each atom has four outer-ring electrons; they form a tight four-corner connection with the four outermost elements of the atom next door. When that happens, all eight electrons are held tightly in place; no electrons can break free, and no current flows. But humans have learned how to fool the silicon atoms by doping the silicon with impurities. They do it by introducing tiny quantities of a different element—arsenic, for example—that has five outer-ring electrons. The four outside electrons of a silicon atom will bind themselves to four of the arsenic electrons—leaving one extra arsenic atom unbound, free to move. The tainted block of silicon is now a conductor. If more arsenic atoms are introduced, more free electrons result and more current will flow. The current moving through the silicon—a flow of free electrons—is the same current that J. J. Thomson saw moving through the vacuum in his cathode ray tube.

But what if the silicon is doped with an element—boron, for example—that has only three electrons in its outermost ring? When silicon atoms try to link up with boron, the arrangement comes up one electron short, leaving a vacant spot—an empty hole—where the eighth electron should be. A nearby electron will be pulled over to fill the hole; this leaves another hole where the electron came from. Another electron moves in to fill this new hole, leaving another hole in its place. The result, effectively, is a movement of holes across the silicon block.

In his classic text on semiconductor physics, William B. Shockley explains this by comparing a block of silicon to a parking lot. In a pure, undoped crystal, every space on the lot is filled and no traffic can flow. If one car is removed, leaving a vacant slot, another car can move ahead, leaving its place open, in turn, for another car to move into. It is the cars that move, of course, but to an observer looking down every once in a while from a high building it appears that the empty space is migrating across the lot. Effectively, Shockley's book says, "the vacant parking place . . . can move owing to the successive motion of vehicles into it."

The doping process results in two different types of silicon. Where the silicon has received extra electrons, it takes on a negative charge, because electrons are negative. Where the silicon is pocked with holes —representing missing electrons—it takes on a net positive charge.

If the phenomenon of holes had been discovered in Thomson's day, when all men of science had a firm grounding in the classics, the vacant spot would most likely have been given a Greek or Latin name, a name like "vacutron" or "nihilon" or some such. The flow of holes was not firmly established, though, until the 1930s, when the classics were in decline and English had become the *lingua franca* of physics. Consequently, the formal scientific name for the positively charged hole is a simple English word—"hole." (Shockley titled his definitive text *Electrons and Holes in Semiconductors.*) With equivalent simplicity, the physicists decided to refer to semiconductor material that had been doped with excess electrons—and is thus negatively charged—as an "N-type" semiconductor. A semiconductor block doped with positive charge—holes—is called "P-type."

By the late 1930s, when all these intricacies had been worked out, the physicists had a reasonably decent picture of what goes on inside a semiconductor. All that remained to put this knowledge to practical use—to launch the semiconductor revolution—was human ingenuity. This essential ingredient was to come from a team of three Americans headed by an intriguing figure who has been, in different seasons, one of the most respected and most reviled of all modern scientists—William B. Shockley.

Shockley, presently an emeritus professor of engineering science at Stanford, was the only child of a technology-oriented couple; his father was a mining engineer, his mother a geologist. Born in London, where his father was stationed, in 1910, he grew up at the northern edge of what is now Silicon Valley. After graduating from Cal Tech he did graduate work at M.I.T., focusing on the movement of electrons in solid materials. Fresh out of school, the young Ph.D. went to work for Bell Labs in 1936. Management, ignoring his graduate work, shunted him off to the vacuum tube department; by sheer persistence, Shockley eventually worked his way into the semiconductor laboratory.

It was a rich, exciting time to be there: semiconductor physics was just falling into place, and there was a sense among the bright, ambitious young men in the field that they were witnessing the dawn of marvelous things. On December 29, 1939, Shockley wrote in his lab notebook: "It has today occurred to me that an amplifier using semiconductors rather than vacuum is in principle possible." Within a decade, he had turned that principle into practice in the form of the transistor, an invention that won Shockley and his colleagues, Walter Brattain and John Bardeen, the Nobel Prize in 1956. Thereafter, with the apotheosis of science that followed *Sputnik,* the mass media made him into a national hero.

In addition to theoretical physics, Shockley has practiced engineering (he holds ninety patents), teaching, and strategic planning (he devised submarine attack methodologies during World War II). His experience in these varied intellectual disciplines has prompted him to do a great deal of thinking about thinking—about the motivations and the thought processes that lead to good ideas. His basic rule for solving any problem is to go back to fundamentals: "Try simplest cases."

Understanding is most likely to result, he says, from reducing the situation to its simplest elements and proceeding from there. His famous book on semiconductors follows the pattern: In Chapter I he sets forth the comparison of atomic structures to parking lots, complete with little sketches of cars to demonstrate traffic flow. By Chapter XV he is explaining that "the wave function A (ø) for the hole-wave packet is not an eigenfunction for the Hamiltonian for the 2N-1 electrons in the valence band."

Motivation is at least as important as method for the serious thinker, Shockley believes, and his scientific papers are replete with asides about the role of this comment or that experiment in spurring him on to new discoveries. The crucial element for successful work in any field, he says, is "the will to think"—a phrase he learned from the nuclear physicist Enrico Fermi in 1940. "In these four words," Shockley wrote later, "[Fermi] distilled the essence of a very significant insight: A competent thinker will be reluctant to commit himself to the effort that tedious and precise thinking demands—he will lack 'the will to think'—unless he has the conviction that something worthwhile will be done with the results of his efforts." The discipline of competent thinking is important throughout life, Shockley says, whether on a prize-winning experiment or a pop quiz in freshman physics. For many years at Stanford he taught a freshman seminar called "Mental Tools for Scientific Thinking"; the basic text was the professor's own essay "THINKING about THINKING improves THINKING."

All this THINKING about THINKING thrust Shockley into a furious intellectual and political controversy which he initially provoked in the late 1960s and which continues, at lower intensity, to this day. In a letter to the National Academy of Sciences, and then in a series of lectures and interviews, Shockley urged detailed study of a problem he named "dysgenics" and defined as "retrogressive evolution through the disproportionate reproduction of the genetically disabled." In plain English, Shockley was claiming that, on the average, blacks are dumber than whites. Thus high birth rates among blacks could lead to "decline of our nation's human quality." "My research leads me inescapably to the opinion," Shockley said, "that the major cause of American Negroes' intellectual and social deficits is hereditary and is racially genetic in origin." Shockley went on to propose a social policy to deal with this situation: government should work to reduce birth rates among low-IQ elements of the population, through programs such as tax breaks for voluntary sterilization.

This theory, proposed by a physicist with no training in genetics and set forth during a fairly tumultuous period in American history, made Shockley one of the most despised and vilified men in the United States. He was denounced as a pseudoscientist, a fanatic, a Fascist. He was burned in effigy on both coasts and denied the right to speak at some of the nation's most prestigious colleges, including Harvard, Dartmouth, and Yale. He was permitted to appear at Princeton, speaking in a small hall (chosen for its tight security) while hundreds of enraged demonstrators screamed protests outside. Back at Stanford, groups of students equipped with bullhorns, the portable and immensely powerful loudspeakers made possible by the transistor, would gather under his office window to chant "Off Pig Shockley"—affording Shockley the experience, probably unique in engineering history, of watching his own invention used to provide hundredfold amplification of demands for his death.

The instigator seemed to relish the stir he had created. A photograph of one of Shockley's classes that has been disrupted by demonstrators in Ku Klux Klan robes and pointed hats shows the professor standing calmly aside, arms folded, taking in the scene with an appearance of benign unconcern. At one of the countless "Off Shockley" rallies at Stanford, the microphone went on the blink; the inventor, who was present, stepped forward and repaired the transistorized device so the speakers could continue their denunciation of his ideas. No matter how bitter the attacks on him, Shockley never stopped seeking new forums to convey his message; in 1982, to the dismay of party leaders, he briefly pursued the Republican nomination for the U.S. Senate. His platform called for a public inquiry into "dysgenics."

One of the ironies of this extended controversy is that Shockley, for all his intensity and determination, seems much more a pleasant grandfatherly type than a public ogre. Those who have worked with him over the years invariably describe him as "charming"; at scientific meetings he was known for telling jokes and even performing magic tricks at the podium while delivering his papers. He also has a reputation for getting the most out of the people who work with him. This he certainly did at Bell Labs when, just after World War II, he was put in charge of a team investigating new semiconductor applications.

The senior member of the group—he was forty-five when the transistor was invented—was Walter Houser Brattain. Brattain grew up on the family ranch in Washington State and went to Whitman College, where he is now an emeritus professor of physics. He did graduate work at Oregon and Minnesota, and upon receiving his Ph.D. in 1929 went to work for the newly organized research institution, Bell Labs. It was a time, as he recalled later, when "the vacuum tube and thermionics were just shedding their baby teeth," and he was first put to work on tubes. Later he moved into semiconductor research under the great scholar C. J. Davisson and was present on the day in 1937 when word arrived that Davisson had won the Nobel Prize. A horde of reporters swarmed onto the premises, and the lab was quickly engulfed in microphones, newsreel cameras, and banks of kleig lights. Davisson, noticing the astonishment on his young assistant's face, stepped over to Brattain and whispered, "Don't worry, Walter—you'll win one someday."

The most unassuming of men, Brattain seems almost embarrassed about winning awards. At the ceremony during which King Gustav VI of Sweden presented him the Nobel Prize, Brattain observed that "one indeed needs to be humble about accepting such an award when he thinks how fortunate he was to be in the right environment at the right time." He is equally diffident about the scientist's role in society. His job, Brattain said, is merely to enhance our understanding of the physical world; "I feel strongly, however, that the scientist has no right to dictate how his understanding is used." Nonetheless, when he took a look back on the twenty-fifth anniversary of the transistor, Brattain did voice one complaint: "The thing I deplore most is the use of solid-state electronics by rock and roll musicians to raise the level of sound to where it is both painful and injurious."

Brattain was the experimentalist of the transistor trio; he had an intuitive feeling for the way semiconductors ought to work, and it was he who turned his colleagues' theories into working apparatus. The first transistor was an ungainly construct of germanium and wire that could be made to work only by Walter Brattain and only, he wrote later, "if I wiggled it just right."

The third member of the transistor team was John Bardeen. Born in 1908, the son of the dean of the medical school at the University of Wisconsin, he grew up in Madison and took his engineering degree at Wisconsin. He went to work for Gulf Oil in Pittsburgh but decided after three years that he preferred pure to applied science. He enrolled at Princeton, studied semiconductor physics under the Nobel laureate Eugene Wigner, and took a Ph.D. in 1936. He taught at Minnesota, worked in the Naval Ordinance Laboratory during the war, and then agreed to join his friend Shockley at Bell Labs in 1946. In 1951, with the basic transistor work completed, he suggested that Bell should undertake work on superconductors. When this proposal was rejected, Bardeen left for a position at the University of Illinois, where he is now a member of the university's Center for Advanced Study. For his superconductor research at Illinois, Bardeen won a second Nobel Prize, in 1972.

Bardeen was a perfect complement to Brattain. His forte is theory. On the transistor project, Brattain would work his experiments and then Bardeen would sift through the results and explain what they meant. He was not the type of person who could wiggle a piece of machinery just right and make it work. On the day in 1972 when his second Nobel was announced, the public relations people at Illinois arranged a press conference. The reporters showed up, but not Bardeen. He apologized later, explaining that he had been stuck at home because he couldn't get his (transistorized) garage door opener to work.

Brattain the experimentalist, Bardeen the theorist, and Shockley, who did some of both, began in 1946 to study the different attributes of the two types of semiconductor: P-type, the material that had been doped with excess positive charges (holes), and N-type, which had excess negative charges (electrons). They worked with "bipolar" germanium—that is, a strip of germanium that had been doped so as to be N-type on one end and P-type at the other.

The men focused intensively on the point in the center where N-type and P-type meet. This point, the most important meeting place in modern physics, is called the P-N junction. The P-N junction makes wonderful things happen.

The P-N junction works like the turnstile you pass through when you enter the subway or a stadium: you can go through easily in one direction but not the other. The P-N junction is a one-way door for electrons; they can pass through it one direction but not the other. When the semiconductor strip is hooked up to a source of current—a battery, for example—electrons can flow easily from the N-type material, across the P-N junction, to the P end. But they can't cross the junction in the other direction.

A device that lets current pass in only one direction—that's a "rectifier," just like John A. Fleming's vacuum tube rectifier, which made reliable radios possible. Since the two ends of the bipolar strip act like the two electrodes of Fleming's tube, the semiconductor rectifier, like the tube, is called a "diode." But instead of the large, hot, fragile, power-hungry vacuum tube that Fleming used, this semiconductor diode, with its minute P-N junction, is tiny, low-power, unheated, and unbreakable. The semiconductor diode needs no vacuum, either; the electronic action takes place within the solid: hence the term "solid state."

By 1946, when the Bell Labs team began its work, the operation of the semiconductor diode was fairly well understood. Semiconductor diodes were not at all important in electronics, however—except for highly specialized applications, like radar—because they did not perform the essential task of amplification. As long as electronic equipment still needed vacuum tubes—with all their attendant problems of high power, heat, and size—as amplifiers, it was just as convenient to use vacuum diode rectifiers as well. The revolution in electronics came when Shockley, Bardeen, and Brattain, after two years of intense and frequently exasperating work, devised a semiconductor triode—a tiny, low-power, unheated, solid-state amplifier.

The world's first solid-state triode was a jury-rigged affair put together by Bardeen and Brattain. It involved two fine wires placed extremely close to each other precisely at the P-N junction on a small piece of germanium. If the wires were set at exactly the right spot—and if Brattain wiggled them just right—a small current flowing into one wire could be amplified to a current one hundred times as great.

The inventors demonstrated this device to the Bell Labs brass on December 23, 1947. At first they measured the amplification with standard electronic meters, but the test that really brought home what had been done came when they hooked up a microphone to one end of their invention and a loudspeaker to the other. One by one, the men picked up the microphone and whispered "hello"; the loudspeaker at the other end of the circuit shouted "HELLO!" Shockley, with an acute sense of history, realized the nice symmetry of the moment: another major breakthrough in electronics had occurred in Bell's lab. "Hearing speech amplified by the transistor," he wrote later, "was in the tradition of Alexander Graham Bell's famous 'Mr. Watson, come here, I want you.' "

This first device, known as a point-contact transistor, was a cumbersome, imprecise instrument and of itself would probably not have made a significant impact. The true importance of that first transistor was that it inspired Shockley to perfect the solid-state amplifier that would revolutionize electronics—the junction transistor. Shockley had only a small role in the development of the point-contact invention. "My elation with the group's success was tempered . . . ," he recalled later. "I experienced some frustration that my personal efforts . . . had not resulted in a significant inventive contribution of my own." This frustration, coupled with the knowledge that solid-state amplification was indeed possible, gave Shockley "the will to think." For the next five weeks, he dedicated himself to "the effort that tedious and precise thinking demands." Working alone on New Year's Eve, he filled nineteen pages in his lab notebook with ideas and sketches. On January 23, 1948, he set down in his notebook the concept of the junction transistor.

Shockley's idea—the basis of all transistors today—was a semiconductor sandwich: a strip of germanium with three different regions—N-type at one end, P-type in the middle, N-type at the other end. This created two P-N junctions, back to back. Three wires were hooked up—one to each N-type region, and one to the P-type in the middle. Now, if a charge was sent to the center region, it would vigorously suck electrons from one N-region, across the center, and out the other end. The stronger the charge on the center region, the more electrons it drove across the junction. That is, the charge in the center region could amplify the flow of current down the strip.

In one common type of transistor, the first N-type region is called the "source"; the center region is the "gate"; the second N-type region is the "drain." Current flows out of the source and down the drain—and the amount of flow is controlled by the gate in the middle. A small

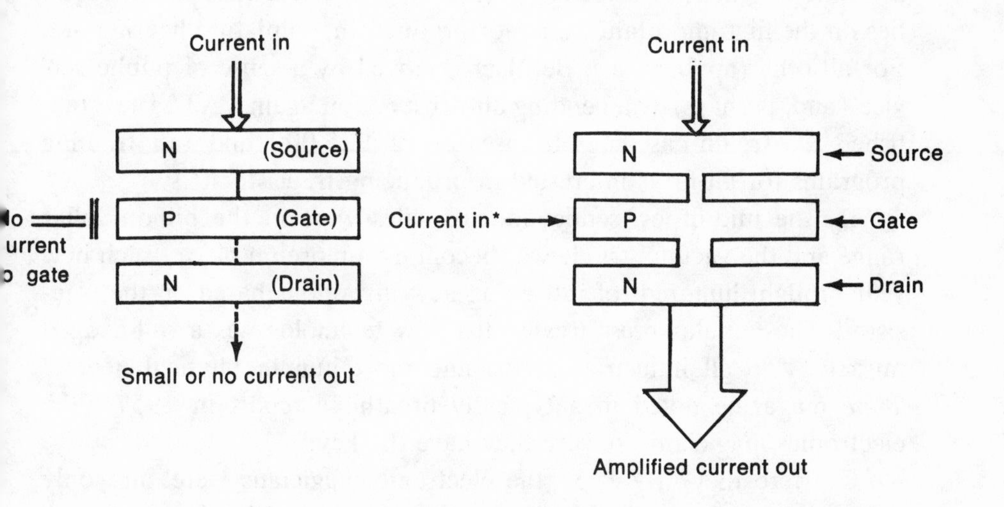

* The greater the current flowing into the gate, the greater the flow from source to drain.

change in the current hooked to the gate causes a big change in the current from source to drain. More important, variations in the gate current are exactly mimicked in the current flowing to the drain.

The three sections of this solid-state sandwich are analogous to the three electrodes of Lee deForest's vacuum triode. The semiconductor "gate" acts like the vacuum tube "grid." A radio signal sent to the gate can shape a much stronger current flowing from source to drain. Just as in the vacuum tube, the effective result is an amplified reproduction of the weak signal. In addition to its application as an amplifier, the solid-state triode can serve as an extremely high-speed switch. If the device is adjusted properly, a signal to the gate will cut off the drain current completely; another signal will open the gate and turn the drain current back on. The important thing is that, in a transistor, this on-off switching takes place in billionths of a second. This was faster than any vacuum tube could be made to switch. This capability made the modern computer possible.

Smaller, lighter, faster, more sensitive, more reliable, and far more power-efficient than the vacuum tube, the transistor could not have been anything other than a stupendous success. Two important decisions by the Bell System accelerated its progress. Mindful of its founder's lifelong interest in helping the deaf, Bell waived all patent royalties on the first important transistor product, the miniature hearing aid. For all other applications, Bell Labs, moved by a sense of public service (and, perhaps, by a pending antitrust action against AT&T), established a bargain-basement license fee of $25,000 and ran training programs for all firms interested in producing transistors.

By the mid-fifties, semiconductor sales were in the billion-dollar range and the vacuum triode was becoming a museum piece. Each new year brought hundreds of intriguing new inventions based on the transistor. The popular press treated the new technology as a full-fledged miracle. "To all industrial needs, and most human physical needs," *Time* magazine noted in a typically breathless report in 1957, "the electronics magicians are sure they have the key."

Or were they? By 1957 the electronic magicians were sure only that they had a serious problem—the problem posed by the tyranny of numbers. Unless the interconnections dilemma could be resolved, the enormous promise of a transistorized future might never be realized. The central importance of the problem—and the profits to be gained from its solution—instilled in governments and research labs and manufacturing concerns around the globe the will to think about a solution, and the will to spend large sums in pursuit of it.

One of the firms in the forefront of this pursuit was Texas Instruments, which mounted a large-scale research project to deal with the numbers problem early in 1958 and began recruiting semiconductor experts from throughout the world for the task. Among the men T.I. hired was a lanky thirty-four-year-old engineer from Milwaukee named Jack Kilby.

3

A Nonobvious Solution

JACK ST. CLAIR KILBY landed his first job in electronics the year the transistor was born and has been working ever since at the front lines of high technology. He was among a small group of pioneers in the 1950s who developed the transistor from a laboratory specimen to a mainstay of industry. He conceived and built the world's first integrated circuit, or semiconductor chip, and then proceeded to invent what is probably the chip's most famous offspring, the handheld calculator. For the past six years he has been working on the photovoltaic effect, an as yet imperfectly understood phenomenon of semiconductor physics, in an effort to construct an electric generating station fueled by the sun. He has won countless scholarly awards, including the National Medal of Science, the technologist's equivalent of the Congressional Medal of Honor. His picture hangs in a hall of fame in Washington between those of Henry Ford and Ernest Lawrence, inventor of the cyclotron.

For all that, however, Jack Kilby in person seems the very antithesis of high tech. Calm and quiet, plain-spoken and plainly dressed (his standard uniform in the lab is an open-necked sports shirt and knockabout cotton trousers), Kilby is the kind of person you might expect to find rocking peacefully on the porch of some country store, his large feet propped up on the railing. He is an imposing figure, not fat but big in every other sense: six feet six inches tall, wide shoulders, massive hands, a large, round, ruddy face framed by a few wayward tufts of gray hair poking up from the temples, and an enormous smile that suggests, accurately, a friendly, casual, unruffled personality. An introvert, he spends a good deal of time alone with his thoughts, working through ideas; he has always done his most creative work on his own. In conversation, he is not quick. Ask him a question—about semiconductors, politics, the best route to the airport—and he will take a long puff on a Carlton, think for a moment in absolute silence, take another puff, and then answer, softly, slowly. The answers are invariably thoughtful and delivered in fully structured sentences that flow perfectly from beginning to end without digression or detour.

Despite his pioneering work in the most modern of technologies, the inventor has an old-fashioned streak. He won't wear a digital watch; characteristically, he has given considerable thought to the difference between digital and analog (i.e., with hands) timepieces and concluded that the older kind better conveys the seamless passage of time. A computer would be useful in his work, but he doesn't use one because "I don't really know how." Although he is probably the single person most responsible for the demise of the slide rule, he still keeps his favorite Keuffel & Esser Log-Log Decitrig handy in the center drawer of his desk, and in some ways he prefers it to the handheld calculators that rendered it obsolete. "It's an elegant tool," he says affectionately. His hobby is photography (black-and-white, of course), for which he has contentedly used the same trusty Hasselblad for thirty years. His car, an aging white Mercedes two-seater, passed the 100,000-mile mark some time back and will probably go double the distance before Kilby thinks about purchasing anything more up to date.

In an industry and a company (he has worked at Texas Instruments on and off for a quarter-century) where "tough" and "aggressive" are terms of high praise, the quiet but friendly inventor is famous as a nice guy. As he treks through the meandering hallways of Texas Instruments' Dallas headquarters—walking with the wary, stooped gait of a man who has bumped his head too often on low ceilings—he greets everybody by name, from top management to messenger boys. Everyone at the company seems to have a story or three about some act of kindness on Kilby's part. To that collection I can add another. When I first called Kilby, out of the blue, to ask if I might spend some time with him, he readily agreed, and then added—it was a first in my journalistic career—that he would pick me up at the airport and shuttle me around because "taxis can be hard as hell to get around here."

Because of the more-or-less parallel development of quantum theory and semiconductor technology, Kilby's work has regularly taken him near or right up to the leading edge of physics. It is a bewilderingly complex field; to understand the flow of charge in a chip, for example, one has to calculate the eigenvalues of the z-component of the angular momentum operator. But Kilby is *not* a scientist. He is quite firm on the point, insisting, in soft but definite tones, that he is an engineer. "There's a pretty key difference," he says. "A scientist is motivated by knowledge; he basically wants to explain something. An engineer's drive is to solve problems, to make something work. . . . That is basically what I have always wanted to do, to solve technical problems. It is quite satisfying, extremely satisfying, to go through the process and find a solution that works."

Kilby has done a great deal of thinking about that process, and, true to form, he has worked out a careful theory of the art of solving a problem, technical or otherwise. Somewhat simplified, the method involves two levels of concentrated thought.

At first, the problem solver has to look things over with a wide-angle lens, hunting down every fact that might conceivably be related to some kind of solution. This involves extensive reading, including all the obvious technical literature but also a broad range of other publications—books, broadsides, newspapers, magazines, speeches, catalogues, whatever happens into view. Kilby himself reads, not skims but reads, two or three newspapers every day and a dozen magazines or so each week. In addition he devours books; his office looks like a publisher's warehouse where the books have staged a coup. The U.S. government issues about 60,000 patents every year, and Kilby tries to read every one. Some of the inventions are arguably related to his work: "anode stud coatings for electrolytic cells"; "electric current regulator." Others seem rather far removed: "gas-fired ceramic radiant poultry brooder"; "dentifrice encapsulation"; "guitar amplifying system"; "self-packaged glider toy"; "golf putting aid"; "3- or 4-product surface-wave acousto-optic time-integrating correlator." "That's all right," Kilby says. "You read everything—that's part of the job. You accumulate all this trivia, and you hope that someday maybe a millionth of it will be useful." For recreation, Kilby says, "I read trash."

The next step in Kilby's system requires switching to an extremely narrow focus, thinking strictly about the problem and tuning out the rest of the world. This requires, first of all, an accurate definition of the problem. "The definition of the problem becomes a major part of the innovation," Kilby has written. "A lot of solutions fail," he says, "because they're solving the wrong problem, and nobody realizes that until the patent is filed and they've built the thing." It is also necessary to develop a clear understanding of the natural constraints surrounding the problem; the heart of the inventor's job is finding a way to slip past the roadblocks erected by nature. "Although invention is considered a creative process," Kilby said once in a lecture on the subject, "it differs appreciably from creativity in the arts. The painter starts with a blank canvas, the author or poet with a blank sheet of paper. They choose an image . . . . and they are free to use any techniques they have to achieve it. Technical creativity is more constrained. The laws of nature, the properties of materials . . . provide very real constraints."

In this concentrated, single-minded focus on the question at hand, the problem solver must also tune out all the obvious solutions. This is a key principle, important to emphasize because it is somewhat counter-intuitive. The mind tends to jump to the answer that is immediately evident. In fact, this answer is probably wrong. If the problem is of any importance, all the obvious solutions have been tried already. The word "nonobvious" appears in few dictionaries, but it is an important part of Kilby's personal lexicon, a concept he returns to again and again when he gets talking about the business of solving problems. Some of history's most important innovations, he says, were so nonobvious as to violate the scientific rules of the day. "You only arrived at the invention when somebody developed a method that everyone else had already decided was obviously wrong."

At this point, if the problem solver's preparation has been broad enough, and he has defined the right problem, and he observes the physical limits, and he's creative, and he's lucky, he might hit on a nonobvious solution that works. But that is not enough, at least not for an engineering problem. The essence of engineering, Kilby says, is cost consciousness. "You could design a nuclear-powered baby bottle warmer," he says, "and it might work, but it's not an engineering solution. It won't make sense in terms of cost. The way my dad always liked to put it was that an engineer could find a way to do for one dollar what everybody else could do for two."

Kilby's dad was an electrical engineer who worked at electric power companies around the Midwest and eventually rose to the presidency of the Kansas Power Company, a medium-sized utility that had small generating plants scattered around the western part of the state and headquarters in Great Bend, a neat, bustling town at the point where the Arkansas River bends south toward the Mississippi. Jack was born in Jefferson City, Missouri, in 1923, but spent most of his childhood in Great Bend. He attended the public schools there, and in the summers he and his father would traverse the plains in a big 1935 Buick, stopping at each of the power company's remote facilities. They would crawl through the works of the generating stations, trying to find out what had gone wrong with a faulty armature or testing the efficiency of a new-model transformer.

When the blizzard of thirty-seven swept through Kansas, blocking roads and felling telephone lines everywhere, the senior Kilby borrowed a neighbor's ham radio to keep track of his far-flung operations. Twisting the dials, turning the big antenna this way and that, squeezing the earphones tight against his head so he could make out the weak, fluctuating signals racing through the Kansas night, Jack was quickly hooked. A brand-new government agency, the Federal Communications Commission, began requiring licenses for radio operators. Jack studied for weeks, took the exam, and came home from school one day to find an official letter assigning him his own set of call letters—W9GTY. He built a ham set, improved it, scavenged some parts, improved it again. By the time he got to Great Bend High School it was clear that he would make his career in electrical engineering, and he set his sights on the engineer's mecca, the Massachusetts Institute of Technology. On a June day in 1941 he boarded the New England States, the crack train connecting the midwestern plains with the great centers of learning and commerce on the East Coast, and rode to Cambridge to take the entrance exam for M.I.T.

At this point, our inventor's story takes a downward turn. Jack failed the test. Forty years later, having launched the Second Industrial Revolution, received more than fifty patents, and won all the leading engineering awards, Kilby still feels the sting of flunking that exam. He can remember his score on the test—he got 497, three points short of passing—and the algebra problems he thinks he got wrong. Back then, however, he didn't really have time to sit around and brood about his fate, because, in addition to the personal blow, the rejection letter from M.I.T. created a practical crisis. Jack had not bothered to apply to any other college. After some scrambling he was admitted to his parents' alma mater, the University of Illinois. He had been there less than four months when the Japanese bombed Pearl Harbor. Freshman Kilby became Corporal Kilby, assigned to a radio repair shop at an Army outpost on a tea plantation in northeastern India.

Everybody learns something in the Army. The eighteen-year-old corporal learned that creative engineering can solve problems that have been officially declared insoluble. His unit was one of the first endeavors of the United States in guerrilla warfare. Small teams of soldiers would be air-lifted over the hump into Burma to put together indigenous resistance movements to harass the Japanese. The Americans kept in touch with their base using radios they carried on their backs. The best portable radios ever made were provided—but these units weighed 60 pounds and broke down regularly under the stress of jungle operations. The Army responded to all complaints by saying that its transmitters, the state of the art in radio, could not be improved. An engineer, of course, knows that there is no machine anywhere that cannot be improved. The engineers in Kilby's unit set up a lab in a dusty pup tent. They sent Corporal Kilby down to Calcutta to buy old radio parts on the black market. Over time, they began turning out ad hoc transmitters that were both lighter and more power-efficient than the official issue.

After the war, Kilby went back to Illinois, eager to learn about radar and other wartime advances in electronics. On the whole, he was disappointed. The one thing he can recall now about his electronics classes was that none of the experiments turned out the way the instructors said they would. There were courses at Illinois on quantum physics and semiconductor phenomena, but they were restricted to scientists; "they weren't going to expose that funny stuff to simple-minded engineers," Jack says. Kilby graduated in 1947 with a traditional engineering education and decent but not outstanding grades. He went to work in Milwaukee at Centralab, for the excellent reason that it was the only firm that offered him a job.

As such things often do, the job turned out to be a perfect spot for the neophyte engineer. Centralab, a division of Globe-Union Corporation, was then producing electric parts for hearing-aid, radio, and television circuits. It was an intensely competitive business, where a cost differential of one dollar per thousand parts—a tenth of a penny per part—could win or lose huge contracts. "It was sort of a crash course in sensitivity to cost," Kilby recalled later. By the late 1940s, radio engineers had already determined the optimum material for each kind of electric component; resistors were made of carbon, capacitors of metal and porcelain, connecting wires of silver or copper. Rather than try to squeeze out a few additional hundredths of a cent per part by improving materials, Centralab was working on the notion that more could be saved by production efficiencies—by placing all the parts of a circuit on a single ceramic base in one manufacturing operation. The firm had only mixed success, but the conceptual seed—that the components of a circuit need not be manufactured separately—was to stay with Kilby and bear important fruit.

In his early years at Centralab, Kilby began to develop his problem-solving methodology. To acquire the wide-angle picture of the problems he had to solve at work, he was determined to learn everything he could about his field. He took graduate courses at night, plowed through the technical literature, attended any lecture that might be interesting. One night at Marquette University he heard a physicist—it was John Bardeen—describe a new invention that achieved amplification and rapid on-off switching *without a vacuum tube*; thereafter, Kilby tried to read everything he could find about this new solid-state device. Then as now, Kilby's reading went far beyond electronics. At one point he happened upon a dental supply catalogue; one page, so unpleasant he can still remember it, described a new technique that used small sandblasters to scour away tooth decay.

Centralab was small and informal enough to let the most junior man in the lab take on important jobs. Kilby was immediately put to work solving real engineering problems, and as he got the hang of it, he acquired a priceless asset for anyone engaged in creative work—confidence. Over time, Jack came to realize that if he approached a problem correctly, worked at it long enough, and refused to let initial failures get him down, he could find a solution. One of the big problems with Centralab's single-step circuit-building process was the reliability of the resistors. The process involved making resistors by printing small patches of carbon on the ceramic base. The machinery was imprecise, so no two carbon patches were the same size. Resistor performance was unpredictable. Kilby was given the job of finding a cheap, simple way to make all the resistors the same size. Starting off with his wide-angle review of the problem, Kilby pondered everything that might be relevant. Something came to mind, something he had read someplace. Those tiny dental sandblasters—did anybody use them? As it happened, the technique had never caught on with dentists, whose patients found it repulsive. But Kilby managed to track down some of those precise devices; this nonobvious approach proved a perfect way to make all the carbon patches exactly the same size.

When Bell Labs announced in 1952 that it would issue licenses for production of its newly patented transistor, Centralab put up the $25,000 license fee and dispatched Kilby to Bell's ten-day crash course in the new technology. There he got a detailed look at the fantastic new world that would be possible now that the limitations imposed by vacuum tubes had been eliminated. He came back to Milwaukee full of ideas, ideas that led to important advances in the manufacture and packaging of transistorized equipment. Gradually, however, he came to realize that the new electronic world had a limit of its own.

Working in a relatively small firm, where the circuit designers in the engineering lab had regular contact with the plant managers, Kilby soon learned—probably sooner than many other people in the business—that realities of the manufacturing process severely restricted the complexity of transistorized circuitry. Kilby and his colleagues upstairs in the lab could design a hearing aid or a radio amplifier that squeezed unheard-of numbers of components into minute spaces. But down in the factory, those circuits could not be built; there were just too many interconnections too close together for the human hand to make them. "Jack Morton at Bell Labs suggested that electronics was facing a 'tyranny of numbers,'" Kilby recalls, "and that was a perfect term for it because the numbers of parts and connections in some of these new circuits were just too big. The simple fact was that you could not do everything that an engineer would want to do."

For Kilby, the recognition of this major new problem was electrifying. Just as he was coming into his own as an engineer, a problem solver, the world of electronics was up against a baffling problem of premier importance. The advent of the transistor offered enormous, world-shaking possibilities—but they would never be realized unless somebody found a way around the problem of numbers. Like everyone else in the industry, Jack Kilby plunged into the search for a solution.

It was evident, though, that solving the tyranny of numbers, if indeed a solution could be found, was a task that would require large resources—considerably larger than a firm the size of Centralab could muster. "I felt," he wrote later, ". . . that it would not be possible for very small groups with limited funding to be competitive. I decided to leave the company." Early in 1958 he sent out his résumé to engineers at a number of larger firms. Among those he wrote to was Willis Adcock of Texas Instruments.

Texas Instruments today, largely because of Jack Kilby, is the world's leading manufacturer of semiconductor devices. In 1958, though, it was just beginning to make a mark in the electronics business. The company had been born in the mid-twenties as the Geophysical Research Corporation; its business was sending sound waves deep into the earth to find potential oil drilling sites. During World War II, the same deep-sounding methods proved useful for locating enemy submarines, and GRC became a defense contractor. The firm's postwar president, Patrick Haggerty, expanded the government business to the point where manufacture of electronic instruments was much more important than geophysical research. Convinced that great things were in store for his little firm, the visionary Haggerty changed the company's name to General Instruments—an audacious suggestion that this impudent pup in Dallas could stand with General Electric and the other eastern electronics giants. The Pentagon didn't like the choice—it had another supplier with a similar name—so Haggerty unhappily fell back on geography: Texas Instruments.

It was another audacious Haggerty gambit that started T.I. on its road to dominance. In 1952, when transistors were still exotic, unreliable devices costing $15 or more each, Haggerty hired a Bell Labs physicist named Gordon Teal and ordered him to develop a reliable mass-production transistor that would sell for $2.50. Teal did it. In 1954, Haggerty launched his most famous initiative: he put his cheap, reliable transistors into a consumer product—the pocket radio. The idea was a smash hit in the marketplace. More important, it made the transistor a common household item and Texas Instruments a common name in electronics.

The first pocket radios, like all transistorized equipment of the day, used transistors made of germanium, a material easy to work with but unsatisfactory for many applications because it could not operate at high temperatures. Another semiconductor material, silicon, could withstand heat but was considered too brittle and too hard to purify for transistor manufacture. Haggerty ordered Gordon Teal and Willis Adcock, a physical chemist, to devise a silicon transistor. The project was pursued under security arrangements that any of the world's spy agencies would admire. In May 1954, Teal and Adcock attended a technical meeting at which speaker after speaker discussed the insuperable problems posed by silicon. Finally, Teal rose to speak. He had listened with interest, he said, to the bleak predictions about silicon's utility. "Our company," he noted calmly, "now has two types of silicon transistor in production. . . . I just happen to have some here in my coat pocket." Adcock then appeared, carrying a record player that employed a germanium transistor. As a record played, Teal dunked the transistor in a vat of boiling oil; the sound stopped. Next Teal wired in one of the silicon transistors. He dumped it into the hot oil; the band played on. The meeting ended in pandemonium. Texas Instruments was on its way.

Soon enough, the tyranny of numbers became evident to the people in Dallas, and Adcock was placed in charge of a major research effort to surmount this obstacle to further progress in electronics. One of the first solutions T.I. worked on was an idea called the "Micro-Module." The theory behind it was that all the components of a circuit could be manufactured in the same size and shape, with wiring built right into each component. These identical modules could then be snapped together, like a child's Lego blocks, to make instant circuits. The concept was important to Texas Instruments not so much because of its intrinsic merits but because it was important to the U.S. Army. Each of the military services was pursuing its own solution to the interconnections problem, and the Army keenly desired that its proposal should prevail; if T.I. could deliver, it would become the darling of all Army contracting officers. Thus when Jack Kilby arrived at Adcock's lab in May of 1958, the Micro-Module was the hottest thing going. Kilby disliked it from the start.

This feeling was partly an engineer's intuition; the Micro-Module bore some resemblance to an idea that had flopped at Centralab, and

Kilby didn't think it would work any better in Dallas. The real flaw, though, was more basic: the Micro-Module implied the wrong definition of the problem. The real problem posed by the tyranny of numbers was numbers. The Micro-Module did nothing to reduce the huge quantities of individual components in sophisticated circuits. No engineer could work with much enthusiasm on a solution to the wrong problem; Kilby's heart sank at the thought that he had left a good job and moved his family across the country only to be put to work on a project that was fundamentally off target.

Texas Instruments then had a mass vacation policy; everybody took off the same few weeks in July. Kilby, who hadn't been around long enough to earn vacation time, was left alone in the semiconductor lab. He was "discouraged," he wrote later; "I felt it likely that I would be put to work on a proposal for the Micro-Module program when vacation was over unless I came up with a good idea very quickly."

Kilby plunged in with his wide-angle approach, soaking up every fact he could about the problem at hand and the ways Texas Instruments might solve it. Among much else, he took a close, analytical look at his new firm and its operations. The obvious fact that emerged was Texas Instruments' heavy commitment to silicon. To capitalize on its victory in the race to develop silicon transistors, T.I. had invested millions of dollars in equipment and techniques to purify silicon and manufacture transistors with it. "If Texas Instruments was going to do something," Kilby explained later, "it probably had to involve silicon."

This conclusion provided Kilby the focus he needed for the narrow, concentrated phase of problem solving. He began to think, and think hard, about silicon. What could you do with silicon?

Jack Kilby's answer to that question has come to be known as The Monolithic Idea. The idea has so changed the world that it is just about impossible today to reconstruct what things were like before he thought of it—and thus it is almost impossible to appreciate how ingenious, and how daring, the answer was. The Monolithic Idea has become an elementary part of modern science, as fundamental, and as obvious, as J. J. Thomson's daring suggestion that there were tiny charged particles swirling around inside the atom. In July of 1958, though, Kilby's answer was hardly elementary.

What could you do with silicon? It was already known in 1958 that the standard semiconductor devices, diodes and transistors, could

Ken Kirkley

After ten years in the engineering boondocks, Jack Kilby (*above*), a quiet, slow-talking sort from Great Bend, Kansas, finally got a chance to work in a major laboratory on a problem of premier importance. Within weeks, he hit on an idea that struck the world of microelectronics like a lightning bolt.

In order to further test the feasibility of circuit on a single crystal wafer, I have had three units made up as shown:

These units were made from diffused material which had been prepared for development work on the 2N623. The wafers were lapped and the slot was etched using a wax mask. Gold plated platinum tabs were alloyed to the back of the wafer. The emitter + base contacts, including those which serve as contacts for the capacitors, were evaporated on before the slot was etched. Plateaus for the transistor and capacitor areas were masked, using wax, and etched. The bar was then cemented to a piece of microscope slide using Sauereisen cement. Gold wire was thermo-bonded to the contact areas, and the other ends of the wire Ecobonded to the gold tabs. These units were fabricated by Tom Yeargan from my sketch.

On testing, it was found that one unit, using a 7 volt supply, would act as a bi-stable device. That is by grounding the base leads alternately,

"It looks pretty crude," Kilby says today when he looks back at the first integrated circuit. "In fact, it *was* pretty crude." But on September 12, 1958, Kilby was able to record in his lab notebook that this crude circuit-on-a-chip had worked perfectly on its first test.

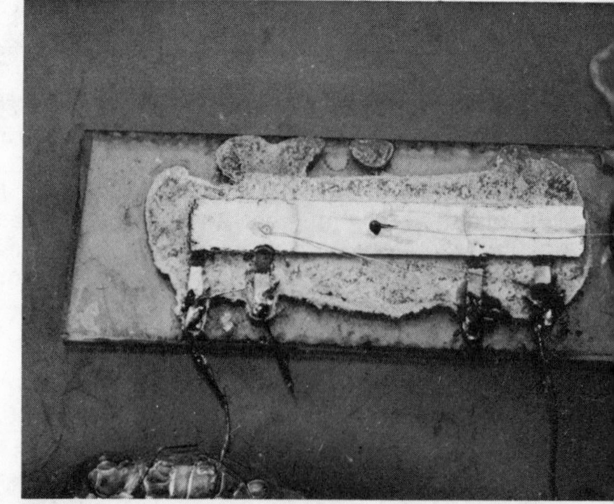

Texas Instruments, Inc.

3,138,743

THE UNITED STATES OF AMERICA

TO ALL TO WHOM THESE PRESENTS SHALL COME:

Whereas Jack S. Kilby, of Dallas, Texas, assignor to Texas Instruments Incorporated, of Dallas, Texas, a corporation of Delaware,

PRESENTED TO THE Commissioner of Patents A PETITION PRAYING FOR THE GRANT OF LETTERS PATENT FOR AN ALLEGED NEW AND USEFUL INVENTION THE TITLE AND A DESCRIPTION OF WHICH ARE CONTAINED IN THE SPECIFICATION OF WHICH A COPY IS HEREUNTO ANNEXED AND MADE A PART HEREOF, AND COMPLIED WITH THE VARIOUS REQUIREMENTS OF LAW IN SUCH CASES MADE AND PROVIDED, AND

Whereas UPON DUE EXAMINATION MADE THE SAID CLAIMANT is ADJUDGED TO BE JUSTLY ENTITLED TO A PATENT UNDER THE LAW.

NOW THEREFORE THESE Letters Patent ARE TO GRANT UNTO THE SAID Texas Instruments Incorporated, its successors OR ASSIGNS FOR THE TERM OF SEVENTEEN YEARS FROM THE DATE OF THIS GRANT

RIGHT TO EXCLUDE OTHERS FROM MAKING, USING OR SELLING THE SAID INVEN- THROUGHOUT THE UNITED STATES.

In testimony whereof I have hereunto set my hand and caused the seal of the Patent Office to be affixed at the City of Washington this twenty-third day of June, in the year of our Lord one thousand nine hundred and sixty-four, and of the Independence of the United States of America the one hundred and eighty-eighth.

Attest:

Attesting Officer.

Commissioner of Patents.

By the time Kilby received "Letters Patent" for his "Miniaturized Electronic Circuits," the right to the invention was already entangled in a labyrinthine legal snarl. The case finally turned on the infamous "flying wire picture." And the winner was . . .

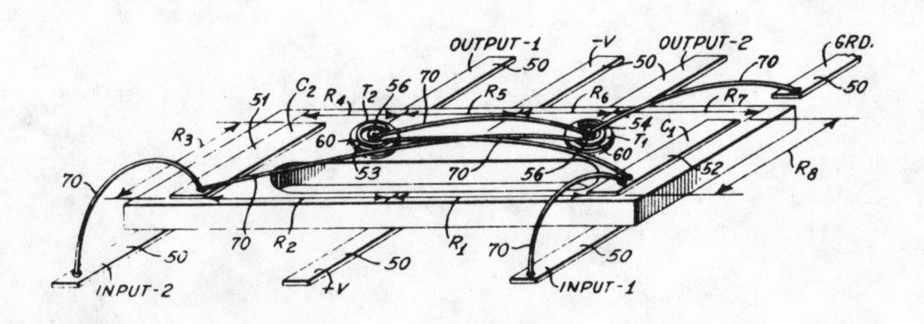

Chuck O'Rear, *Time* magazine

. . . Bob Noyce, the minister's son from Denmark, Iowa, a brilliant, multifaceted physicist and manager who hit on The Monolithic Idea independently. Noyce's first integrated circuit was based on the "planar" process, which eliminated the flying wires and made the chip a practical reality.

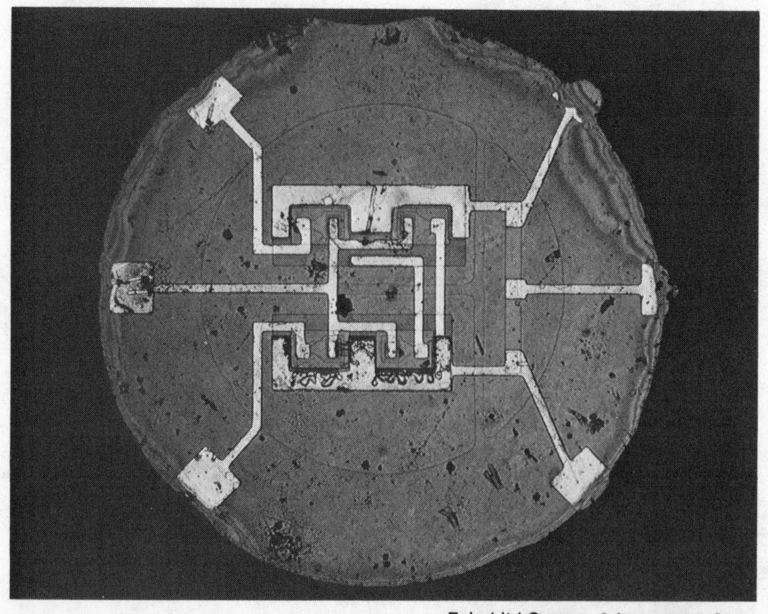

Fairchild Camera & Instrument Corp.

A.T.&T. Bell Laboratories

Kilby and Noyce are spiritual descendants of another American genius, Thomas A. Edison, who first noticed the phenomenon of thermionic emission and built the "Edison Effect" lamp—a light bulb with a long glass nose sticking out—to study it.

It was the estimable J. J. Thomson (*left*), successor to Clerk Maxwell's chair at Cambridge, who finally explained what the Edison Effect was all about. In the process, J.J. discovered the electron and opened the door to a whole new scientific cosmos known as subatomic physics.

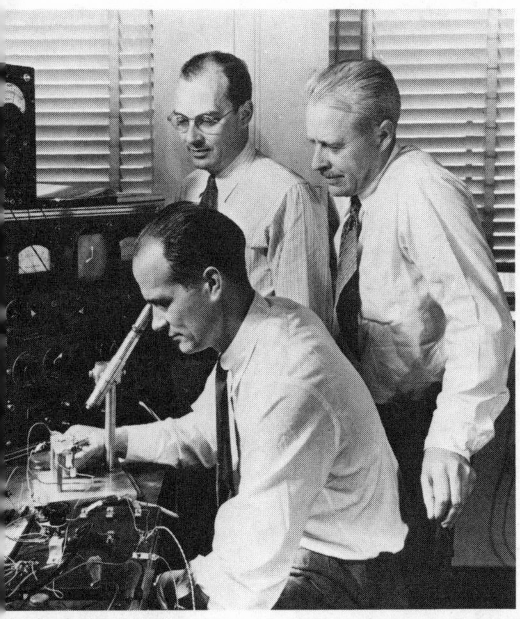

National Park Service

A half-century after J.J.'s discovery, William Shockley (*seated*), John Bardeen (*standing, left*), and Walter Brattain eliminated the glass bulb and put electrons to work in the seminal invention of the semiconductor age: the transistor.

Texas Instruments, Inc.

Patrick Haggerty (*above, left*) had the revolutionary notion that he could put the transistor into every home—by building the "transistor radio." The big radio makers scoffed at this hare-brained idea, but Haggerty persuaded Edward Tutor (*above, right*), of tiny Regency Radio, to give it a try.

Ten years later, Haggerty decided he could put the microchip into every household as well. He told Jack Kilby to invent a portable adding machine. The result was the first pocket calculator (in the model's right hand, *below*). Today's little calculators make Kilby's breakthrough look like a dinosaur.

Texas Instruments, Inc.

A.T.&T. Bell Laboratories

The Microelectronics Revolution: from tube (*left*) to transistor (*center*) to microchip (*right*).

Apple Computer, Inc.

Apple Computer, Inc.

The first digital computers, like the University of Pennsylvania's ENIAC (*above*), filled large rooms and consumed the power of a locomotive to heat their thousands of vacuum tubes. By the year 2000, today's chip-based nonpareil, the lap computer (*right*), will probably look as oversized as ENIAC does today.

A.T.&T. Bell Laboratories

The 32-bit microprocessor packs the computing power of 10,000 ENIACS on a flake of silicon thinner than a hair and smaller than an infant's fingernail. "The technology is still far from the fundamental limits imposed by the laws of nature," Bob Noyce wrote. "Further miniaturization is less likely to be limited by the laws of physics than by the laws of economics."

be made of silicon if the silicon was doped with the proper impurities to make it conduct electric charges. But if the silicon had no impurities, its electrons would all be bound in place. No charges would flow through such a piece of silicon; it would block current just like a standard resistor. Kilby thought about that: a silicon resistor? Why not? A strip of undoped silicon could act as a resistor. It wouldn't be as good as a standard carbon resistor, but it would work. For that matter, by taking advantage of the peculiarities of the P-N junction, you could also make a capacitor out of silicon. Not much of a capacitor—its performance wouldn't equal that of a standard metal-and-porcelain capacitor—but it would work. And come to think of it—this was the idea that would revolutionize electronics—if you could make all parts of a circuit out of one material, you could manufacture all of them, all at once, in a monolithic block of that material.

The more Kilby thought about it, the more appealing this notion became. If all the parts were integrated on a single slice of silicon, you wouldn't have to wire anything together. Connections could be laid down internally within the semiconductor chip; no matter how complex the circuit was, nobody would have to solder anything together. The numbers barrier would disappear. And without wiring or connections, an awful lot of components could be squeezed into a pretty small chip. On July 24, 1958, Kilby opened his lab notebook and wrote down The Monolithic Idea: "The following circuit elements could be made on a single slice: resistors, capacitor, distributed capacitor, transistor." He made rough sketches of how each of the components could be realized by proper arrangements of N-type and P-type semiconductor material.

This suggestion was, in a word, nonobvious. "Nobody would have made these components out of semiconductor material then," Kilby has explained. "It didn't make very good resistors or capacitors, and semiconductor materials were considered incredibly expensive. To make a one-cent carbon resistor from good-quality semiconductor seemed foolish." Building a resistor out of silicon seemed about as sensible as building a boxcar out of gold; you could probably do it, but why bother? Even Kilby was a little skeptical at first: "You couldn't be sure that there weren't some real flaws in the scheme somewhere." The only way to find out was to build a model of this integrated circuit and give it a test. To do that, Kilby would need the boss's okay.

When everybody came back from vacation, eager to get cracking on the Micro-Module, Kilby showed his notebook sketches to Willis Adcock. "Willis was not as high on it as I was," Kilby recalled later. Adcock was intrigued with the idea but had doubts about its practicality; "it was pretty damn cumbersome," he said later. It was probably worth trying—but on the other hand, Adcock was supposed to be making Micro-Modules. To build Kilby's model, he would have to divert people from that project and put them on the previously untried task of building a complete circuit out of semiconductors. Eventually, Kilby and Adcock made a deal: if Kilby could make a working resistor and a working capacitor out of separate pieces of silicon, Adcock would authorize the far more costly effort to construct an integrated circuit on a single semiconductor chip. Kilby painstakingly carved a resistor out of a strip of pure silicon. Then he took a bipolar strip of silicon and wired the P-N junction to make the capacitor. He wired these strange devices into a test circuit, and they worked. Adcock then okayed the attempt to construct a complete circuit on a single chip.

The design that Adcock chose was a "phase-shift oscillator" circuit, a classic unit for testing purposes because it involves all four of the standard circuit components. An oscillator is the opposite of a rectifier; it turns direct current into alternating current. If it works, the oscillator transforms a steady, direct current into fluctuating pulses of power that constantly change direction, back and forth, back and forth. The transformation shows up neatly on an oscilloscope, a piece of test equipment that displays electric currents graphically on a television screen. If you hook direct current—for example, from a battery—to the oscilloscope, the steady current will show up as a straight line across the screen, like this:

But if you put a phase-shift oscillator between the battery and the oscilloscope, the oscillating current will show up as a gracefully curving line—a sine wave—undulating across the screen, like this:

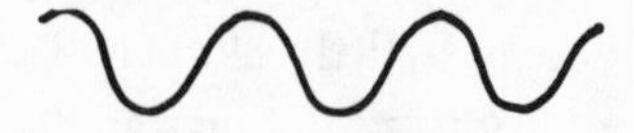

On September 12, 1958, Jack Kilby's oscillator-on-a-chip, half an inch long and narrower than a toothpick, was finally ready. A group of Texas Instruments executives gathered in Kilby's area in the lab to see if this tiny and wholly new species of circuit would really work. Conceptually, of course, Kilby knew it would; he had thought the thing through so often, there couldn't be a flaw. Or could there? After all, nobody had ever done anything like this before. Kilby was strangely nervous as he hooked up the wires from the battery to his small monolithic circuit, and from the circuit to the oscilloscope. He fiddled with the dials on the oscilloscope. He checked the connections. He looked up at Adcock, who gave him a here-goes-nothin' shrug. He checked the connections again. He took a deep breath. He pushed the switch. Immediately a bright green snake of light started undulating across the screen in a perfect, unending sine wave. The integrated circuit, the answer to the tyranny of numbers, had worked. The men in the room looked at the sine wave, looked at Kilby, looked at the chip, looked at the sine wave again. Then everybody broke into broad smiles. A new era in electronics had been born.

4

Leap of Insight

THE LITTLE AIRPLANE jumped from the boy's hand and shot off into the blue Iowa sky, the engine purring, the body spiraling perfectly, the plane racing higher and farther away every second, soaring right past the end of town and far out over the cornfields. It was almost a mile away, still performing beautifully, when he lost sight of it for good. "That was my first technological disaster," Robert Norton Noyce recalled many years later. "Yeah, it worked a lot better than any model I'd ever built before, but it was gone." Actually, it wasn't gone; six months later, during the harvest, a farmer found the toy plane among the corn stalks and guessed that it was probably the work of the minister's son, who was always messing around with engines and gadgets and models.

By then, though, the boy had moved on to other things. That lost airplane had prompted him to build a radio control unit for his next model. Radio proved so interesting that Bob Noyce and a buddy put together a pair of crude transceivers to send messages back and forth. Neither boy obtained a radio operator's license, making their network a federal offense; the crime can safely be reported now because the statute of limitations for their violation expired about 1942. The boys got interested in chemistry; soon Bob was mixing his own home-brew explosives, gun cotton and nitrogen tri-iodide. Then he found an old washing machine motor and tried to make it drive his bicycle. Then there was something else, and something else after that. "I was a pretty curious kid," Noyce says now. "I was always trying to figure out how everything worked."

Some things never change. The most striking thing today about Robert N. Noyce is the enormous range of his interests and the breadth of his activities. He still wants to know how everything works. At various stages of his professional career he has been a theoretical physicist, an inventor, a corporate chief executive, a venture capitalist, a lobbyist. In recent years he has emerged as the elder statesman and leading spokesman for the American semiconductor industry—"the mayor of Silicon Valley," as a colleague puts it. He has been successful, extraordinarily successful, both in science—he has won a slew of academic awards and holds what has been legally declared the primary patent for the integrated circuit—and in business. Estimates of his net worth today range upward from $100 million. Meanwhile, off the job, he restores old airplanes and helps plan Harvard's future and skis and studies Japanese and takes up whatever else his far-flung curiosity fastens on.

An affable, naturally gregarious person, Noyce seems to gravitate to a leadership position in most things he undertakes. He got interested in madrigal singing some time ago, joined a chorus, and eventually became its conductor. A friend once took him to an Audubon Society meeting; soon enough, Noyce was overseeing an effort to find a new habitat for a dwindling species of Newfoundland puffin. He was one member of a large group of Silicon Valley executives who decided a few years back that the microelectronics business needed a professional organization; soon enough, Noyce was chairman of the Semiconductor Industry Association.

A wiry, athletic-looking man with angular features and curly hair that is starting to gray, Noyce conveys the rakish self-assurance of a racing driver or jet pilot (despite a clear memory of that first disaster, Noyce is still fascinated with airplanes and flies a rebuilt Republic Seabee, a World War II seaplane). A fashionable dresser, he wears silver-rimmed aviator glasses, a large gold ring, and a platinum watch that has both old-style hands and new-style digital readout. If Jack Kilby seems as loose as a spare piece of string, Noyce is a coiled spring of energy and enthusiasm. When the *Harvard Business Review* asked him to describe the employees of his present corporation, Intel, Noyce responded with an answer that might serve as a self-portrait. "They're high achievers," he said. "High achievers love to be measured, when you really come down to it, because otherwise they can't prove to themselves that they're achieving." Among such people, Noyce added, "I don't think you could call it a relaxed atmosphere. A confident environment, but not a relaxed one."

One reason it is difficult to relax at Intel is that the firm deliberately keeps itself on the leading edge of technology, always looking for the new idea. This is the way Robert Noyce likes it; he is a "technologist," he says, which means "the kind of person who is comfortable with risk." That's the key difference between a technologist and businessman, Noyce says: "No businessman would have developed the telephone. It's got to be a maverick—some guy who's been working with the deaf and gets the crazy idea that you could actually send the human voice over a wire. . . . A businessman would have been out taking a market survey, and since it was a nonexistent product, he would have proven conclusively that the market for a telephone was zero."

The same description—"comfortable with risk"—applies to Noyce's approach to technical questions. He attacks engineering problems with absolute confidence that there is a perfectly good solution somewhere and that the human mind can find it. He tends to be a prolific, impulsive producer of solutions himself, but he also knows that "a lot of my great ideas turn out to be ridiculous when you look closely." For that reason, Noyce is not a solitary inventor; he needs to work with others, to talk things over, to launch his ideas and see if they will fly. "You explain a lot of things to yourself," he says, "by trying to explain them to someone else. And then either you can see for yourself that the idea won't work, or the other person can spot the problem and help you find a better way."

Early in his career Noyce came into the ken of another famous technical optimist, William B. Shockley, and it is clear that he absorbed Shockley's first rule for solving problems: "Try simplest cases." "All of our progress, in any technical field, well, there are a few leaps of insight, but beyond that it has been decomposing it to its simplest elements, to understandable elements, and building it back up from its simplest element," Noyce says. "Shockley had this wonderful ability to make the right simplifying assumption, to get the math out of the way until you had a basic visual image of what was happening. . . . Well, you see, you have to get back to the basic, the simplest picture before you can understand a problem well enough to solve it."

Like Kilby, Noyce tends to be instantly suspicious of any solution that seems too obvious. "It's absolutely true, whether it's a technical question or anything else, that a lot of people are going to try to approach a problem the same way everybody else has," Noyce says. "You can just let yourself get in a rut. And if you don't get out of the rut, you're not going to solve the problem." This is where the "leap of insight" comes in. "The successful solution comes about because somebody was able to fire up his imagination and try something new. . . . If you want to achieve something worthwhile, you have to jump to the new idea."

Noyce spent his entire boyhood pursuing new ideas, new phenomena, new gadgets. He was born in 1927 in the tiny town of Denmark, Iowa, the son of a Congregationalist minister. The family moved from one small rural town to the next when Bob was small; when his father took over the parish in Grinnell, a quiet, attractive town fifty miles east of Des Moines, the family settled there for good. Although the elder Noyces had no particular penchant for technology, they encouraged their sons and listened with interest each night as Bob and his three brothers explained their latest experiments. To pay for his airplane kits, radio parts, chemicals, etc., Bob worked as a baby-sitter and mower of lawns. One of his chief customers was Prof. Grant Gale, chairman of the physics department at Grinnell College. Under Gale's tutelage, Noyce fell in love with math and physics. He studied the college texts while in high school, and when he enrolled at Grinnell in 1945 he knew from the first that physics and math would be his major interests. Not his only interests, of course—that was not the way Bob Noyce lived. He was the star diver on Grinnell's swimming team, he sang in choral groups, played oboe in a band, and had a continuing role in a radio soap opera.

Noyce's life story is basically a narrative of one outstanding success after another, but at college he experienced an unforgettable setback. One night in 1948 a group of Grinnell boys decided—it evidently seemed logical at the time—that what their dormitory really needed was a genuine Hawaiian luau, complete with roast whole suckling pig. Tasks were divvied up: Bob and another athletic sort were assigned the job of acquiring the pig—a mission they accomplished by swiping a suckling out of the sty at a nearby farm. The Iowan luau was a great success, but next morning, when the repentant Noyce confessed his crime, nobody was laughing. In Iowa, theft of livestock was not a humorous matter. Thanks to his father's stature and Grant Gale's intervention, he was spared a criminal sentence and was kicked out of college for only one semester. He spent the time working at an insurance company. When he reentered Grinnell midway through his senior year he was still able to graduate with his class, earning top grades and making Phi Beta Kappa.

For the most part, the electronics Noyce learned in his physics classes at Grinnell was standard vacuum tube stuff—the Fleming rectifier, the deForest amplifier, and various improvements to those basic devices. One day, however, Professor Gale astounded the class with news of something totally different. Gale had been a classmate of John Bardeen in the engineering school at the University of Wisconsin, and thus he was able to obtain one of the first transistors and demonstrate it to his students. "It was simply astonishing," Noyce remembers. "Just the whole concept, that you could get amplification without a vacuum. It hit me like the atom bomb. It was one of those ideas that just jolts you out of the rut, gets you thinking in a different way."

Fascinated by the new solid-state technology, Noyce applied to M.I.T. and was admitted to a graduate course in physical electronics. He took his Ph.D. there in 1953, sifted through a long list of job offers, and went to work in Philadelphia for Philco, which was just embarking on a large-scale effort to develop and produce improved transistors. The job gave Noyce a chance to practice serious science. He turned out a series of monographs and papers on semiconductor devices. In one paper, Noyce reported on the effects of low-energy gas discharges on a platinum-germanium diode; in another, he demonstrated that the "dc transistor current amplification factor" could be determined using this formula:

$$\alpha = \text{Sech}\ (W_b/L_b)\ \{1 + [(J_{rq'} + J_{d'})/J_d\ \text{Tanh}\ (W_b/L_b)]^{-1}$$

Late in 1955, Noyce gave a paper before the American Physical Society on "base widening punch-through." That paper caught the eye of the nation's preeminent semiconductor specialist.

On a January day in 1956, Noyce received a telephone call from

William B. Shockley. Shockley explained that he was leaving Bell Labs and moving to California to start a new company that would develop high-performance transistors. Would Noyce be interested in interviewing for a job? "It was like picking up the phone and talking to God," Noyce recalled later. "He was absolutely the most important person in semiconductor electronics. Getting that job meant you would definitely be playing in the big leagues." Noyce took a cross-country train to Palo Alto. With characteristic confidence, he spent the morning of his arrival renting a house near Shockley's lab; that done, he went to the interview to see if he could land the job. He got it, settled his family into the new home, and set to work with Shockley on the development of a high-performance double-diffusion transistor.

The transistor, as Shockley conceived it, is a semiconductor sandwich in which a thin region of P-type silicon (that is, silicon that has been doped with impurities so that it contains extra positive charges) is sandwiched between two regions of N-type silicon (silicon doped with excess negative charges). (There are also less common transistors that employ the opposite structure, with N-type material in the center and P-type on either end.) To get the best performance characteristics from the device, the separate regions of the silicon chip should be clearly defined, and the P-N junction, the point where the different regions meet, should be a sharp, sudden transition from P-type to N-type material. Through the early 1950s transistor makers tried dozens of different techniques to achieve the precise doping and the sharply defined junctions required for reliable transistor action. The process that eventually proved best—the process still used today in semiconductor manufacture—was a Bell Labs discovery called "diffusion."

Semiconductor diffusion works like a barbecue pit where hickory smoke seeps into the meat and imparts a distinctive flavor. In the diffusion process, a bar of silicon is cooked in a furnace at high heat, and then a gas containing the appropriate doping impurities—boron, for example, or arsenic—is pumped into the furnace. At temperatures of 1000 degrees centigrade or so, some of the impurity atoms in the gas seep into the silicon bar and "dope" it, either with excess electrons or with positively charged holes. Just as a barbecue chef knows just how long to cook the ribs to get the right taste of hickory, solid-state physicists gradually determined the proper time and temperature needed to put precise amounts of impurities at precise points on the silicon block.

Thus diffusion provided the first effective means of creating a semiconductor bar with sharply defined regions of N- and P-type silicon.

The process used to make transistors was called "double diffusion." A round wafer of silicon, about the size of a 45 rpm record in diameter and about the thickness of five pages of this book, would be placed in the furnace. Two different kinds of impurities would be diffused onto the wafer in separate steps, leaving a three-layer cake—N-type on the bottom, P-type in the middle, N-type on top. The wafer could then be cut, like a cake, into dozens or scores of tiny three-layer pieces—each one an N-P-N transistor.

Double diffusion made possible, for the first time, the mass production of precise, high-performance transistors. The technique promised to be highly profitable for any organization that could master its technical intricacies. Shockley therefore quit Bell Labs and, with financial backing from Arnold Beckman, president of a prestigious maker of scientific instruments, started a company to produce double-diffusion transistors. The inventor recruited the best young minds he could find, including Noyce, Gordon Moore, a physical chemist from Johns Hopkins, and Jean Hoerni, a Swiss-born physicist whose strength was in theory. Already thinking about human intelligence, Shockley made each of his recruits take a battery of psychological tests. The results described Noyce as an introvert, a conclusion that should have told Shockley something about the value of such tests. Early in 1956, Shockley Transistor Laboratories opened for business in the sunny valley south of Palo Alto. It was the first electronics firm in Silicon Valley.

In Robert Noyce's office today there hangs a black-and-white photo that shows a jovial crew of young scientists offering a champagne toast to the smiling William Shockley. The picture was taken on November 1, 1956, a few hours after the news of Shockley's Nobel Prize had reached Palo Alto. By the time that happy picture was taken, however, Shockley Semiconductor Laboratories was a chaotic and thoroughly unhappy place. For all his technical expertise, Shockley had proven to be an inexpert businessman. He was continually shifting his researchers from one job to another; he couldn't seem to make up his mind what, if anything, the company was trying to produce.

"There was a group that worked for Shockley that was pretty unhappy," Noyce recalled many years later. "And that group went to Beckman and said, hey, this isn't working. . . . About that time, Shockley got his Nobel Prize. And Beckman was sort of between the devil and the deep blue sea. He couldn't fire Shockley, who had just gotten this great international honor, but he had to change the management or else everyone else would leave."

Confused and frustrated, eight of the young scientists, including Noyce, Moore, and Hoerni, decided to look for another place to work. That first group—Shockley called them "the traitorous eight"—turned out to be pioneers, for they established a pattern that has been followed time and again in Silicon Valley ever since. They decided to offer themselves as a team to whichever employer made the best offer. Word of this unusual proposal reached an investment banker in New York, who offered a counterproposal: instead of working for somebody else, the eight scientists should start their own firm. The banker knew of an investor who would provide the backing: the Fairchild Camera and Instrument Company, which had been looking hard for an entrée to the transistor business. A deal was struck: each of the eight young scientists put up $500 in earnest money, the corporate angel put up all the rest, and early in 1957 the Fairchild Semiconductor Corporation opened for business a mile or so down the road from Shockley's operation.

Noyce recalls that the group had some slight qualms about running their own business, but these doubts were easily overcome by "the realization, for the first time, that you had a chance at making more money than you ever dreamed of." The dream, as it happened, came true. In 1968 the founders sold their share of Fairchild Semiconductor back to the parent company; Noyce's proceeds—the return on his initial $500 investment—came to $250,000. Noyce and his friend Gordon Moore had by then found another financial backer and started a new firm, Intel Corporation (the name is a play on both *Intel*ligence and *Int*egrated *El*ectronics). Intel started out making chips for computer memories, a business that took off like a rocket. Intel's shares were traded publicly for the first time in 1971—on the same day, coincidentally, that Playboy Enterprises went public. On that first day, stock in the two firms was about equally priced; a year later, Intel's shares were worth more than twice as much as Playboy's. "Wall Street has spoken," an investment analyst observed. "It's memories over mammaries." Today, Intel is a billion-dollar company and Noyce's shares are worth considerably more than $50 million.

The men who started Fairchild Semiconductor in 1957 were determined to make the double-diffusion transistor, but they were also on the outlook for any other product that could turn a profit. Noyce, as director of research and development, was the man responsible for spotting important technical problems that Fairchild might profitably solve. The most important problem by far facing electronics then was the tyranny of numbers. Off and on, all through 1957 and 1958, Noyce thought about the interconnections problem. In retrospect, he can see now, The Monolithic Idea should have come to him much earlier. "Here we were in a factory that was making all these transistors in a perfect array on a single wafer," Noyce says, "and then we cut them apart into tiny pieces and had to hire thousands of women with tweezers to pick them up and try to wire them together. It just seemed so stupid. It's expensive, it's unreliable, it clearly limits the complexity of the circuits you can build. It was an acute problem. The answer, of course, was don't cut them apart in the first place—but nobody realized that then." Instead, Noyce was stuck in a rut. He worked on standard ideas for making circuit components in smaller sizes and uniform shapes. He came up with nothing worthwhile.

Unable to get out of the rut, Noyce turned his attention to another technical dilemma. Double-diffusion transistors—the tiny three-layer chips of N-P-N silicon—were highly susceptible to contamination. A piece of dust, a stray electric charge, a minute whiff of contaminating gas would break down the P-N junctions and impair transistor action. One day in 1958, Jean Hoerni came to Noyce with a theoretical solution: he would place a layer of silicon oxide on top of the N-P-N chip, like icing atop the three-layer cake. The oxide would hold fast to the silicon and protect it from contaminants.

"It's building a transistor inside a cocoon of silicon dioxide," Noyce explains, "so that it never gets contaminated. It's like setting up your jungle operating room. You put the patient inside a plastic bag and you operate inside of that, and you don't have all the flies of the jungle sitting on the wound."

The people at Fairchild recognized Hoerni's idea—it was called the "planar process" because it left a flat plane of oxide atop the silicon—as an important advance in transistor technology. Noyce quickly called in the firm's patent lawyer, John Ralls, to put together a patent application. Ralls, sensing that this planar idea might have other applications in electronics, wanted to write the application in the broadest language. He told Noyce that Intel should make the application as expansive as he possibly could. Every time they talked about it, Ralls would pose a challenge: "What else can you do with this idea?"

Looking back today, Noyce can see clearly that it was the lawyer's question that pushed him out of his mental rut and provoked the leap of insight that became The Monolithic Idea. What else? What else could you do? In the first weeks of 1959, Noyce was thinking hard about that question, scratching pictures in his notebook, talking things over hour after hour with his sedate, cautious friend Gordon Moore.

As Noyce looked over Hoerni's "planar" idea, he realized that it had another useful property. It was quite difficult in those days to make precise electrical connections to the separate regions of an N-P-N transistor, because wires were relatively large compared to the tiny regions of the chip. Hoerni's oxide icing, spread atop the three-layer silicon cake, helped solve this problem. Connecting wires could be poked down through the icing to the exact spot on the chip, and the oxide would keep them firmly in place. "Remember, what I'm trying to do is make this [transistor] extremely small," Noyce explained recently. "Well, I can't attach a wire to that, because it's too small. But now, with the planar coating, I can attach a big old wire—big old wire, this is, you know, a quarter of a human hair—running it on top of the oxide."

But that realization led to a new idea, something that was even better: wires wouldn't be needed at all. Noyce now saw that tiny lines of copper or some other metal could be printed onto the oxide layer, and thus all the transistor's interconnections could be made at once in a single manufacturing process. And if you could connect the separate regions of a single transistor with these printed metal lines, then you could put two separate transistors on a single piece of silicon and connect them with the printed lines. And if you could put two transistors on a single chip of silicon, couldn't you build some other circuit components on the same chip? Couldn't you, in fact, build a complete circuit, an *integrated* circuit, all on a single chip of silicon? Wouldn't that overcome the tyranny of numbers?

"I don't remember any time when a light bulb went off and the whole thing was there," Noyce says. "It was more like, every day, you would say, well, if I could do this, then maybe I could do that, and that would let me do this, and eventually you had the concept."

One day Noyce walked into Moore's office and showed him, on the blackboard, that two transistors in a single silicon block could be connected by printed copper lines on the oxide layer. A few days later, he was back at the blackboard, showing Moore how he could use a channel of undoped silicon in the same block as a resistor. A few days later, he was drawing a silicon capacitor on the blackboard. It was all completely new, but Moore raised no serious objections.

On January 23, 1959, "all the bits and pieces came together," and Noyce filled four pages of his lab notebook with a remarkably complete description of an integrated circuit. "In many applications," he wrote, "it would be desirable to make multiple devices on a single piece of silicon, in order to be able to make interconnections between devices as part of the manufacturing process, and thus reduce size, weight, etc. as well as cost per active element." Noyce went on to explain how resistors and capacitors could be fabricated on a silicon chip, and how the whole monolithic circuit could be connected by metal contacts printed right onto the chip. He also set forth a rough sketch of a computer circuit—a circuit that would add two numbers—realized in integrated form. Six months after Jack Kilby had reached the same destination, Bob Noyce had arrived at The Monolithic Idea.

News travels quickly in the electronics industry. By the spring of 1959, rumors about a major new development at Texas Instruments had reached the people at Fairchild. Nobody knew exactly what T.I. had done, but it was not impossible to guess which problem this breakthrough was designed to solve. Noyce again called in John Ralls and asked the lawyer to prepare a patent application for a new idea—a "unitary circuit structure . . . to facilitate the inclusion of numerous semiconductor devices within a single body of material." This time, Ralls decided to write a detailed, precise patent application, a document that could serve as a shield to protect Fairchild against any possible legal action by Texas Instruments. This strategic decision would become the decisive factor in a bitter ten-year legal battle fought all the way to the United States Supreme Court.

5

Kilby v. Noyce

THE TERRIFYING RUMOR that raced through the semiconductor lab at Texas Instruments on the morning of January 28, 1959, turned out, eventually, to be wrong on almost every count, but like many false alarms it had the salutory effect of scaring people into action. More than four months had passed since Jack Kilby had successfully demonstrated his prototype integrated circuit, but since then further development of the concept had been almost nil. Kilby's superiors were hoping to introduce their great new product in March, but as of the end of January, there really wasn't any product. The only integrated circuits in existence were the crude models Kilby had built by hand for his demonstration; nobody had figured out yet how to turn out a production version. Even the lawyers were behind schedule; they had failed to take the most basic steps to protect Texas Instruments' right to the new invention.

That's why the rumor was so frightening. At a technical meeting, somebody from T.I. thought he heard somebody else saying that he had been told that another company had come up with an integrated semiconductor circuit that would solve the tyranny of numbers. There was, in fact, a germ of truth in this report; just five days earlier, Bob Noyce, in his office at Fairchild, had scratched his first sketchy concept of The Monolithic Idea in his notebook. But the rumor that reached Dallas had nothing to do with Noyce. The word at T.I. was that somebody at RCA had come up with an integrated circuit, and that—even worse—RCA was soon going to file for a patent. When this unsettling news reached Samuel M. Mims, T.I.'s senior staff lawyer, Mims didn't hesitate a second before putting through an emergency call to Mo Mosher.

Ellsworth H. Mosher, a partner in the Washington, D.C., patent law firm of Stevens, Davis, Miller & Mosher (cable address: "INVENTION") and an elder statesman of the patent bar, knew immediately that he had to act fast. He dispatched a junior lawyer to Dallas and told Mims to sit down with Kilby and find out precisely what the inventor thought his Monolithic Idea would be good for. It normally took two or three months to put together all the paperwork, prose, and pictures required for a patent application; in this case, though, Mosher promised to deliver a completed application to the patent office within a week.

Mosher's advice—that Texas Instruments had better apply for a patent, and fast—was not quite so obvious as it might appear. One of the most important rules that patent lawyers try to get across to their clients is that, in some cases, it is better not to apply for a patent at all. For an inventor a patent is a sort of Faustian bargain.

The patent expressly guarantees the inventor "the right to exclude others from making, using, or selling" the idea for the seventeen-year life of the patent. The patent holder can, if he chooses, issue licenses to others to make, use, or sell the idea, and license fees can bring in large sums of money. If anybody tries to market the patented product without obtaining a license, the inventor can go into federal court to get an injunction and money damages. In exchange for those benefits, though, the patent holder has to reveal all the secrets of his success. The patent law says that an inventor must provide "a written description of the invention, and of the manner and process of making and using it, in . . . full, clear, concise and exact terms." The inventor and his company might have expended a dozen years and a hundred million dollars perfecting the idea; once a patent is granted, anybody in the world can acquire the plans—full, clear, concise, and exact—from the Patent Office for 50 cents.

If, for example, John S. Pemberton had applied for a patent for the formula he whipped up in his backyard one day in 1886, the product that he invented—a soft drink which he named "Coca-Cola"—would have entered the public domain in 1903, when the patent expired, and anybody in the world would have been free from that day forward to brew and sell the drink without paying a penny to the Coca-Cola Company. But Pemberton kept his formula unpatented, and thus secret. Even without a patent, Coca-Cola has been able to defend its formula under a body of law known as "trade secret protection," which makes it illegal to copy deliberately somebody else's commercial idea.

From the inventor's viewpoint, the flaw with the trade secret laws is that they apply only to purposeful stealing of an idea; they do not prevent anybody from marketing a product that he has invented on his own, even if an earlier inventor has been selling the same product for years. Lacking a patent, Coca-Cola would have no recourse against a company selling exactly the same drink if the second firm could prove in court that its chemists had been messing around with sugar, flavorings, and cola nuts and just happened to hit on the precise formula that Coca-Cola uses. The holder of a patent, in contrast, can go to court to stop any competitor from selling the same product, even if the competitor developed the product completely on his own. The strategic decision facing every inventor, then, is whether he wants seventeen years of the stronger protection provided by a patent, or permanent protection under the trade secret laws against only those who deliberately steal the idea.

The choice has to be made because an inventor can take either patent protection or trade secret protection but not both. This principle was established once and for all in the landmark case of Kellogg v. Nabisco—the great Shredded Wheat decision. The familiar shredded wheat biscuit was invented in 1895 by a Colorado baker named Henry Perky, who promptly—and foolishly, as it turned out—took out a patent (No. 548,086) on his new breakfast cereal. The National Biscuit Company subsequently bought the rights to Perky's invention. Despite extensive advertising, shredded wheat sales never took off until the 1920s—long after the patent had expired. Somebody in Battle Creek realized that the formula was free for the taking, and grocers everywhere began carrying a new product—Kellogg's Shredded Wheat. Nabisco went to court, claiming, under the trade secret laws that Kellogg had deliberately copied its product. The shredded wheat litigation wound its way for years through the legal system, and in 1938 it finally reached the United States Supreme Court. In a decision by Louis

Brandeis, the Court sided with Kellogg all the way. Once the patent had been issued, Brandeis wrote, Nabisco lost its right to claim that shredded wheat was its private secret. The basic goal of the patent system, after all, was to encourage public disclosure of technological advances.

This goal was so important to the Founding Fathers that the Patent Office (the name comes from the Latin verb *pateo,* "to open") was one of the earliest federal agencies created by the First Congress. Among the small group of men who constituted the U.S. government then, jobs were assigned largely on the basis of personal interest. The Secretary of State, Thomas Jefferson, was an inventor, so Congress gave the Patent Office to the State Department. In the evening, after a long day of diplomacy, Jefferson would review patent applications. Among the patents he personally granted was one to Eli Whitney in 1794 for the cotton gin.

Inevitably, bureaucracy reared its head. In 1804 the Patent Office hired an employee of its own, and that proved to be the first step along a steep slope; today the Patent and Trademark Office (now a wing of the Commerce Department) has about 3,000 employees scattered around three huge buildings in a suburban mall in Arlington, Virginia. The government receives about 110,000 patent applications each year and grants about 60,000 patents. The office is, among other things, one of the world's busiest publishing houses; it keeps every one of the 4½ million patents issued since Jefferson's time in print and sells about 13,000 individual copies every day.

Like most traditional institutions, the Patent Office has developed parlance and procedures all its own. There are tens of thousands of pages of statutes, regulations, guidelines, and legal opinions governing the issuance of patents. Among much else, the regulations go on for fifteen long paragraphs describing the kind of paper and ink an applicant should use for drawings of his invention ("The sheets may be provided with two ¼-inch (6.4 mm) diameter holes having their centerlines spaced 11⁄16 inch (17.5 mm) below the top edge and 2¾ inches (7.0 cm) apart, said holes being equally spaced from the respective side edges").

At the core, though, the basic rules governing what kind of inventions will be granted a patent are straightforward. The invention has to be "new." It has to be "useful," a term the courts have interpreted to mean that the gadget has to work. The Patent Office receives a few applications each year from people who have invented perpetual motion machines; it rejects them all on the ground that they can't do what they're supposed to and so aren't useful. The inventions also have to be desirable; the government refuses to grant patents for nuclear weapons, no matter how new or how "useful" they might be.

When Mo Mosher heard about the important new invention at Texas Instruments, he knew immediately that T.I. would need a patent. Unlike soft drinks and breakfast cereals, electronic gear rarely has a market life longer than the seventeen-year term of a patent, so it is almost always a wise move for an electronics inventor to seek a patent. Mosher knew, too, that none of the statutory requirements that govern patentability would pose a problem. Kilby's monolithic circuit was clearly something completely new, and since it promised a solution to the most important problem facing the industry, it was eminently useful and desirable.

Before he could start writing Kilby's application, though, Mosher had to resolve a fundamental tactical question. Anyone who applies for a patent has to decide whether he needs it for offensive or for defensive purposes—whether, to use the lawyers' favorite metaphor, he wants his patent to be a sword or a shield. The decision usually turns on the novelty of the invention. If somebody has a genuinely revolutionary idea, a breakthrough that his competitors are almost sure to copy, his lawyers will write a patent application they can use as a sword; they will describe the invention in such broad and encompassing terms that they can take it into court for an injunction against any competitor who tries to sell a product that is even remotely related. In contrast, an inventor whose idea is basically an extension of or improvement on an earlier idea needs a patent application that will work as a shield—a defense against legal action by the sword wielders. Such a defensive patent is usually written in much narrower terms, emphasizing a specific improvement or a particular application of the idea that is not covered clearly in earlier patents.

Probably the most famous sword in the history of the patent system was the sweeping application filed February 14, 1876, by a teacher and part-time inventor named Alexander Graham Bell. That first telephone patent (No. 174,465) was so broad and inclusive that it became the cornerstone—after Bell and his partners had fought some 600 lawsuits against scores of competitors—of the largest corporate family in the world. In the nature of things, though, few inventions are so completely new that they don't build on something from the past. The majority of patent applications, therefore, are written as shields—as "improvements" on some earlier invention. Some of the most important patents in American history fall into this category, including No. 586,193, "New and Useful Improvements in Transmitting Electrical Impulses," granted to Guglielmo Marconi in 1898; No. 621,195, "Improvements in and Relating to Navigable Balloons," granted to Ferdinand Zeppelin in 1899; No. 686,046, "New and Useful Improvements in Motor Carriages," granted to Henry Ford in 1901; and No. 821,393, "New and Useful Improvements in Flying Machines," granted to Orville and Wilbur Wright in 1906.

The integrated circuit, however, was so new, and so potentially lucrative, that Mosher, Mims, and Kilby decided to shoot the works—to write a far-reaching patent application that Texas Instruments could wield as an offensive weapon against anyone else who tried to make, use, or sell the device. Accordingly, Mosher and his partners set out to write a document that would leave no doubt about the revolutionary quality of Kilby's idea.

They laid it on thick. Every aspect of the integrated circuit was described as "novel" or "unique" or both. "Radically departing from the teachings of the art," the application said, ". . . the present invention has resulted from a new and totally different concept for miniaturization." As a result "the ultimate in circuit miniaturization is attained." And for good measure: ". . . the invention . . . represents a remarkable improvement over the prior art."

The document went on to describe two specific circuits that could be fabricated on a semiconductor chip. But Mosher emphasized that Kilby's invention could be used to build an integrated version of any circuit: "There is no limit upon the complexity or configuration of circuits that can be made in this manner." To put a point on his sword, Mosher added a thinly veiled warning to anybody who might try to circumvent the patent by making small changes: "Such changes and modifications," the application said, "are deemed to fall within the purview of this invention."

For all this bravado, the authors of Kilby's patent application still had to contend with a fairly sticky problem. Since the patent system is designed to let the world know the secrets behind technological advances (and thus, in theory, stimulate new advances), the law requires an inventor to explain in his patent application how the device is made. "The specification," the statute says, ". . . shall set forth the best mode contemplated by the inventor for carrying out his invention." Further, the law requires an inventor to provide drawings showing precisely what his new device looks like.

This was a problem because nobody at Texas Instruments, not even Jack Kilby, had figured out the best mode for carrying out the invention. And nobody knew what a production-model integrated circuit would look like.

In a way, Kilby's invention was almost too ingenious, too violent a break with the way things had always been done before. By the late 1950s, engineers had been building electric circuits for almost 100 years, and they had always done it the same way: taking individual resistors, capacitors, and other components and wiring them together. The invention of the transistor had led to dramatic changes in the size, cost, reliability, and efficiency of circuits, but it had not changed the basic structure of discrete components wired together. That was one reason the tyranny of numbers had proved such a baffling problem; to the engineers, a large number of discrete parts was what circuits were all about.

The Monolithic Idea—the idea that an entire circuit could be a single part, with resistors, capacitors, diodes, and transistors built right in—involved two fundamentally new concepts, which Kilby later defined as "integration" and "interconnection." Integration was the central idea: the realization that all the parts of a circuit could be made from the same material—silicon—and thus a whole circuit could be integrated in a single silicon chip. Interconnection was the recognition that, once all the parts of a circuit were made on a single chip, the connections between the different parts—the wires—could be printed onto the chip as part of the production process. That way, the enormous cost and the inherent unreliability of hand-wiring huge numbers of components to one another would be eliminated.

In the first chips he built, during late summer and fall of 1958, Kilby successfully integrated all the components of circuit into a silicon chip. But he didn't have time to work out the interconnections within the chip. Racing to demonstrate the new idea to his bosses before they lost interest, he had settled for a jury-rigged device in which the separate components on the chip were connected (by hand) with tiny gold wires, giving that first integrated circuit the appearance of an intricate cobweb spun by a golden spider. This delicate, handmade circuit was adequate for its purpose—proving that a circuit on a chip would really work—but the job of attaching the wires was much too painstaking and time-consuming for any large-scale production. Before the integrated circuit could become a practical solution to the numbers barrier, somebody would have to deal with the question of interconnection.

By the beginning of 1959, Kilby had somewhat improved his techniques for constructing components within a chip, and he and some colleagues had worked out several reliable designs for different circuits on a chip. But the interconnections business was still up in the air. Kilby had a vague idea, an idea that was similar to the "planar" process that Noyce's colleague, Jean Hoerni, had developed at Fairchild. He thought he could put a layer of silicon oxide on top of the chip—like icing on a cake—and then print fine lines of a conducting metal, such as gold, atop the icing to connect various parts of the circuit. (This was, in fact, the method eventually developed for chip production.) But at the end of January, the only chips in existence still used the handmade cobwebs of wire.

Here was a dilemma: Worried that RCA was going to apply for a patent any minute, the T.I. people were anxious to file first. But a patent application required a picture. To draw a picture, Mosher's artist needed a model. But the only model available was Kilby's crude demonstration chip, with its network of gold wire. After a week or so of deliberation, it was decided to put speed first; the application was filed with a picture of the hand-wired chip:

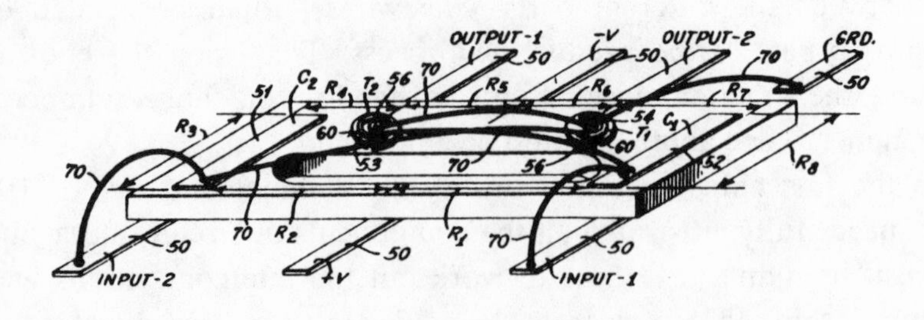

Even now, a quarter-century later, engineers tend to smirk when they see this drawing—it is known in the electronics business as the "flying wire picture"—because it is so far removed from what an integrated circuit is supposed to be.

Jack Kilby, of course, was one of the engineers who knew that the flying-wire drawing was fundamentally wrong. Accordingly, the lawyers threw in some language designed to fudge the issue. "Although the invention has been shown and described in terms of specific embodiments," they wrote, "it will be evident that changes and modifications are possible which do not in fact depart from the inventive concepts taught herein."

That helped a little, but Kilby still wasn't satisfied; he had to put in something to show that his revolutionary new circuit would not have to be wired by hand. At the last minute, accordingly, Kilby and Mims added one more paragraph to the application. "Instead of using the gold wires in making electrical connections," it said, "connections may be provided in other ways. For example . . . silicon oxide may be evaporated onto the semiconductor circuit wafer. . . . Material such as gold may then be laid down on the [oxide] to make the necessary electrical connections." With this final addition, the application—four pages of pictures and five of text—was delivered to the patent office on February 6, 1959.

There, Kilby's application for a patent on his "Miniaturized Electronic Circuits" was assigned to an examiner who specialized in inventions involving electronic circuits and devices. The patent office employs about 1,000 examiners (they are generally lawyers with a background in some technical field) whose job is simple to describe—they have to determine whether an invention is legally entitled to a U.S. patent—but often extremely complicated to perform.

The examiner has to determine whether an idea is "new"—a question that requires searching dozens, hundreds, or thousands of earlier patents, reports, and monographs. He has to decide whether it is "useful," a determination that demands extensive research in the technical literature. And he has to determine whether the inventor's application clearly sets forth how the new gadget is to be built. On the average, each examiner is assigned between 75 and 100 patent applications every year. As a result, it regularly takes months, and sometimes years, for an examiner to deliver his initial opinion on an application. If the examiner finds anything amiss—and he almost always does—he writes a letter to the inventor explaining his reservations. The inventor gets up to six months to prepare a reply or amend his application; then the examiner may take months more to review the reply. It is not uncommon—particularly for a complex invention in a specialized field—for an inventor to wait years for final action on a patent application.

So it was for Kilby. The process dragged on, with the examiner raising questions and Texas Instruments doing its best to answer them. The first important development came some twenty-six months after the application had been filed. On April 26, 1961, Kilby received a telephone call from the lawyers in Washington, informing him that the first patent ever for an integrated circuit had been granted.

But it had not been granted to Jack Kilby.

Like their counterparts at Texas Instruments, the people at Fairchild Semiconductor took their sweet time at first about developing The Monolithic Idea into a practical integrated circuit. It was nearly two months after Bob Noyce had set down the basic concept in his notebook before Fairchild started working on the idea; another four months passed before Noyce got around to filing for a patent.

The reason, Noyce explained later, was that, after twenty months of preliminary planning and organizing, Fairchild at the start of 1959 was just beginning to sell its first important product—the double-diffuse transistor. "We were still a brand new company," Noyce recalled. "We were worried about basic survival. That meant getting transistors out the door. The integrated circuit seemed interesting, it was something that might make you some money somewhere down the road, but that was not a period when you had a lot of time for it." During the winter of 1959, The Monolithic Idea was literally put on a shelf; except for his friend and sounding board, Gordon Moore, Noyce didn't even mention it to any of his colleagues.

Like their counterparts at Texas Instruments, the people at Fairchild were eventually prodded into action by a rumor—although in this case, the rumor was true. Sometime in late February or early March, word arrived in Silicon Valley that Texas Instruments was about to announce a wholly new kind of circuit that would do away with discrete electric components by integrating all the parts of a circuit into a single silicon chip. For any company that made its money selling discrete components—like Fairchild's transistors—this was disconcerting news. Somebody at Fairchild called a meeting to discuss the development. At this session Noyce explained for the first time that he, too, had worked out the basic concept of an integrated circuit. The group decided immediately to get hopping on the new product.

One of the first things to be done was to file for a patent. Noyce and his patent lawyer, John Ralls, did not have access to Kilby's patent application; the patent office treats pending applications as classified material, and there are no leaks. But by mid-March, when T.I. publicly announced its new "Solid Circuit," they knew that the Dallas firm must have filed already. What Fairchild needed, then, was a legal shield—a patent that would differentiate Noyce's version of the idea from the Kilby invention and thus permit Fairchild to enter the integrated circuit market without fear of legal action by T.I.

In preparing his application, Noyce had one significant advantage. Unlike Kilby, who had been forced to file for a patent before he had worked out the problem of interconnections within the chip, Noyce's formulation of the idea covered both integration and interconnection. This was a result of the different routes the two inventors had taken to arrive at the integrated circuit. Kilby had first hit upon the concept of integration—of building all the parts of a circuit in a monolithic chip of silicon—and had moved from there to consideration of interconnections. Noyce, in contrast, had first recognized the possibility of printing connecting strips of metal on a chip—something made possible by Jean Hoerni's invention of the planar process—and the notion of interconnection had lead him to the idea of integration. By the spring of 1959, Fairchild was busily engaged in working out the details of Hoerni's planar process, and thus Noyce had no difficulty providing a description, and a drawing, of a chip with interconnections built right in.

Accordingly, Noyce and Ralls titled their application "Semiconductor Device-and-Lead Structure" ("lead" is the electricians' term for a connecting wire in a circuit), and they strongly emphasized the interconnections aspect of the Noyce circuit. The application listed three "principal objects" of the invention. The first one was interconnections: "to provide improved device-and-lead structures for making electrical connections to the various semiconductor regions."

Noyce's application went on to describe a "unitary circuit structure" that would permit integrating "more than one circuit device into a single body of semiconductor." "According to prior practice," it continued, "electrical connection . . . had to be made by fastening wires directly to the [components]. . . . By means of the present invention, the leads can be deposited at the same time and in the same manner as the [components] themselves."

To the four-page written description of Noyce's idea, Ralls added three pages of pictures of typical circuits that could be integrated onto a chip. There were no flying wires; for that matter, there were no wires at all. The drawings show a structure that is essentially the same as the integrated circuits being produced today.

It was midsummer before Ralls and Noyce were satisfied that everything was in order; the patent application was finally filed on July 30, 1959. It was assigned, evidently, to an examiner who was not aware of the earlier application from Texas Instruments, and it moved ahead at what is, for the patent office, lightning speed. Twenty-one months after it was filed, Noyce's application was granted: U.S. Patent No. 2,981,877. By formal decree of the United States of America, Robert N. Noyce—the second person to come up with the idea—had been officially declared the inventor of the integrated circuit.

The award of an integrated circuit patent to Noyce evoked consternation, but not outright panic, at Texas Instruments. Kilby and his lawyers, after all, were veterans of the patent game; they knew that some applications move through the Patent Office faster than others, and that it is not particularly unusual for the second version of an invention to be the first patented. This happens so often, in fact, that the government has a special procedure—it is called an "Interference Proceeding"—and a special board—the Board of Patent Interferences—to consider the claims of inventors who find themselves in Kilby's position. The basic rule governing an Interference is that priority prevails—i.e., whichever inventor can prove to have had the idea first gets the patent.

Mosher filed the necessary papers, and in May of 1962 both Noyce and Kilby received a copy of Commerce Department Form POL-102, declaring that the Board of Patent Interferences had convened Interference No. 92,842, "*Kilby* v. *Noyce*," to determine who had really been the first to invent the integrated circuit. The board enclosed a short form asking each man to list the earliest date on which he could prove he had had the idea. Since both Kilby and Noyce had maintained lab notebooks precisely for this purpose, both were able to provide a precise answer: July of 1958 for Kilby, January of 1959 for Noyce. With these preliminaries concluded—they consumed about ten months—the stage was set for a final determination.

Actually, it was not quite set. Before the central legal battle could get underway, the lawyers fought out a series of preliminary skirmishes:

"Motion to Dissolve Under Rule 232(a)(2)"
"Opposition to Motion to Dissolve Under Rule 232(a)(2)"
"Motion to Dissolve Under Rule 232(a)(3)"
"Opposition to Motion to Dissolve Under Rule 232(a)(3)"
"Request to File Affidavits"
"Opposition to Motion to File Affidavits"

While this was going on, the Patent Office concluded that Jack Kilby's initial application was satisfactory after all. In June of 1964, Kilby was granted a patent—No. 3,138,743—for the integrated circuit. This made the Interference proceeding even more crucial, and the lawyers went back to work:

"Motion for Extension of Time"
"Opposition to Motion for Extension of Time"

Each of these preliminary disputes took a few months to resolve. And so it was not until July 28, 1964—more than two years after the Interference was begun—that the inventors and their lawyers gathered in Mosher's Washington office to hear the first piece of evidence in the case.

The session was a brief one, directed strictly to proof that Kilby had made the invention first. Kilby explained that he had gotten the idea in July of 1958, and one of Mosher's associates related how he had filed Kilby's patent application in February of 1959. There was a short cross-examination, and everybody went home.

Three months later, everyone met again in a lab in Palo Alto to hear Fairchild's response. There was not a great deal the Fairchild people could say on the basic question, priority of invention. On that point, Kilby was a clear winner. Nonetheless, before the session ended, Noyce's lawyers had dropped a bombshell.

During the long months of preliminary backing and filing, Fairchild's trusted patent attorney, John Ralls, had died. His place was taken by Roger Borovoy, a junior member of Ralls's firm who had caught Noyce's eye and eventually won his confidence. A lot of young lawyers might have found Borovoy's position somewhat daunting. In his first big case he was litigating against Mosher, a titan of the bar, and the facts on the crucial issue were all on Mosher's side. As Borovoy saw it, however, he was sitting pretty. "Here I was," he recalled later, clearly savoring the memory, "a punk kid, defending the most important electronics patent in twenty-five years and Mo Mosher opposing me. Fantastic, right?"

Since he had no reasonable defense to Kilby's claim of priority, Borovoy decided to go on the offensive. He pored over Kilby's patent application, looking for a weak spot to attack. Almost immediately, he found one: the flying wire picture. By 1964, when Borovoy took over the case, the industry had largely determined what an integrated circuit would look like; it didn't look anything like the drawing in Kilby's patent application. Focusing on that drawing, Borovoy drew up a plan of attack. If he could discredit Kilby's application because of its weakness on interconnections, Noyce would be left with the only valid patent for an integrated circuit.

Of course, Kilby's application also contained that last-minute paragraph explaining how, in place of the flying wires, "conducting material such as gold may then be laid down on the insulating [oxide] to make . . . connections." To win the case, Borovoy knew, he would have to find something wrong with that paragraph.

When the litigants assembled at Palo Alto to hear the Fairchild testimony, accordingly, Borovoy brought forward an expert witness—an electrical engineering professor at Stanford—who declared that no one could build an integrated circuit by following the instructions in Kilby's application. The hand-wired circuit in the picture was obviously wrong. For that matter, the business about laying down gold on an oxide was faulty, too. You can lay down gold on oxide, the expert testified, but "it will not stick."

Under Borovoy's gentle prodding, the expert contrasted Kilby's language—"laid down on"—with the wording of Noyce's patent, which said the connection material had to be "adherent to" the oxide layer. "Laid down on" had no clear meaning, the expert said. "Adherent to," in contrast, was a precise technical term. On that fine distinction Fairchild would have to rest its case.

A month later, when inventors and lawyers gathered again at a lab in Dallas to hear Texas Instruments' rebuttal, Mosher produced an expert of his own. This was an engineer from Kilby's alma mater, the University of Illinois, and he thoroughly disagreed with the Stanford man. Gold *will* stick to an oxide layer, he said, so there was no practical difference between "laid down on" and "adherent to." With this testimony, both sides had had their say. All the expert testimony had consumed six more months, but the stage was set for final resolution.

Actually, it was not quite set. First, there were a few more procedural battles to be fought:

"Request for Sur-Rebuttal Testimony"
"Opposition to Request for Sur-Rebuttal Testimony and Conditional Request for Sur-Sur-Rebuttal Testimony"
"Reply to Request for Suspension of Action on Request for Leave to Take Sur-Rebuttal Testimony and on Conditional Motion to Take Sur-Sur-Rebuttal Testimony"

Next, the lawyers had to argue the case before the Board of Patent Interferences. In oral argument and in their written briefs, both sides gave most of their attention to the interconnections question. Borovoy's brief included an oversized copy of the flying-wire picture. "Note that this drawing shows no oxide layer and no gold wires 'laid down on' any such layer," the brief said. "In fact, it is readily apparent that the gold wires are anything but 'laid down.' "

Six months later, however, on February 24, 1967, when the board issued its opinion, it brushed all that aside. After reviewing the experts' disagreement over "laid down on" and "adherent to," the board observed that "we are not particularly impressed with that testimony." As the board saw it, Kilby's patent application, while not perfect, was clear enough. That left only the question of which inventor was first: "Since Noyce took no testimony to establish any date prior . . . Kilby must prevail." Seven years after he had filed his patent application,

Jack Kilby had been adjudicated the inventor of the integrated circuit. The stage was now set for Texas Instruments to wield its sword against the rest of the electronics industry.

Actually, it was not quite set. Any American who is unhappy with a federal agency's decision has the right to appeal, and Fairchild exercised the right. A year was devoted to the preparation of briefs and the filing of motions, and in the fall of 1968, Mosher and Borovoy appeared before the Court of Customs and Patent Appeals to argue all the issues once again. Another year passed. On November 6, 1969, the court issued its opinion.

The decision dealt exclusively with the difference between "laid down on" and "adherent to." The judges had found Roger Borovoy's argument appetizing—and swallowed it whole. "Kilby has not demonstrated," the opinion said, "that the term 'laid down' had . . . or has since acquired a meaning in electronic or semiconductor arts which necessarily connotes adherence." In ignoring the difference between the crucial phrases, the appeals court said, the Interference Board was "clearly in error." The board's opinion was reversed. The Borovoy ploy had worked.

Now it was Mosher's turn to appeal. Six months after the opinion, he filed a brief in the Supreme Court, asking the justices to review the opinion. Six months later, the Court issued a terse reply to Mosher's request: "Denied." Ten years and ten months after Jack Kilby had first applied for his patent, the case of *Kilby* v. *Noyce* had come to an end. Noyce had won. Now the stage was set for Fairchild to exploit its patent.

Actually, though, the stage was not set. During the decade that the lawyers had been waging their battle, the integrated circuit had emerged as the most important new product in the history of electronics. The market grew explosively. By the time the last court had issued the last ruling, production of semiconductor chips was a multi-billion-dollar industry. As a result, the legal right to this invention had become too important to be left to lawyers.

And so, in the summer of 1966, before the first opinion was issued, executives from Texas Instruments, Fairchild, and about a dozen other electronics firms had held a summit meeting—and cut a deal. T.I. and Fairchild each conceded that the other had some right to the invention. The two companies agreed to grant licenses to each other for integrated circuit production. Any other firm that wanted to enter the market then had to arrange separate licenses with both Texas Instruments and Fairchild. The two firms generally demanded a royalty fee ranging from 2 to 4 percent of the licensee's profit from chip production. This agreement provided the other firms a means to enter the integrated circuit business; it provided T.I. and Fairchild with more than $100 million in royalties over the years.

The scientific community, meanwhile, agreed to agree that Kilby and Noyce deserved joint credit for The Monolithic Idea. The two men were both awarded the National Medal of Science for overcoming the tyranny of numbers, and both were inducted into the National Inventors' Hall of Fame. Today, Kilby is generous in describing Noyce's work on the invention, and Noyce is equally generous about Kilby. In the textbooks, Kilby gets credit for the idea of integrating components on a chip, and Noyce for working out a practical way to connect those components. Among their fellow engineers, Kilby and Noyce are referred to as "co-inventors" of the chip, a term that both men find satisfactory.

On the day of Fairchild's great victory, consequently, hardly anybody paid any attention. After ten years and tens of thousands of pages and well over a million dollars in legal fees, the legal labors had brought forth an utterly inconsequential mouse. "Patent Appeals Court Finds for Noyce on IC's," began the headline over a small story in the trade journal *Electronics News* reporting the decision: "IC Patent Reversal Won't Change Much."

The Real Miracle

THE INTEGRATED CIRCUIT made its debut before electronic society at the New York Coliseum on March 24, 1959. The occasion was the industry's most important yearly get-together—the annual convention of the Institute of Radio Engineers. Texas Instruments had managed, in the nick of time, to turn out a few chips that had no flying wires, and there was a lavish display at the T.I. booth featuring the new "Solid Circuits." There was also a lavish prediction (which we know today to have been accurate) from T.I.'s president, who said that Jack Kilby's invention would prove to be the most important and must lucrative technological development since the silicon transistor. Nonetheless, the new circuit-on-a-chip received a frosty reception.

"It wasn't a sensation," Kilby recalls dryly. There were about 17,000 electronic products on display at the convention (the Coliseum used a million watts of power daily during the gathering), and large numbers of them attracted more attention than the integrated circuit. In its special issue on the convention, *Electronics* magazine offered breathless reports on such innovations as a backward-wave oscillator and a gallium arsenide diode but made no mention of the integrated circuit. In a wrap-up two weeks later, *Electronics* devoted a single paragraph to Texas Instruments' new "match-head size solid-state circuit."

"There was a lot of flack at first," Kilby recalls, and indeed, what little comment the new device received was largely critical. The critics identified three basic problems with the integrated circuit. In the first place, the idea of making resistors and capacitors out of silicon flew in the face of decades of research which had established conclusively that nichrome was the optimum material for making resistors, Mylar for capacitors. Monolithic circuits of silicon would be inherently inferior. In the second place, integrated circuits would be hard to make; one common line of analysis held that 90 percent of each production batch of chips would be faulty. In the third place, the whole concept posed a threat to an important segment of the engineering community. If component manufacturers like Texas Instruments started selling complete circuits to computer manufacturers, the circuit designers at computer firms would become redundant—and unemployed.

"These objections were difficult to overcome," Kilby wrote later, "because they were all true." As a result, the giants of the industry—Sylvania, Westinghouse, and their ilk—carefully kept themselves clear of the business for several years. This untimely burst of caution opened the way for upstarts like Texas Instruments, Fairchild, and a slew of new firms in Silicon Valley to work out the problems and cash in on the revolution. With intensive research, these hungry young companies learned how to design circuits on the chips that circumvented the shortcomings of silicon components; they found new production techniques that overcame the initial manufacturing difficulties. The result has been American industry's greatest postwar triumph. The integrated circuit, a child of Texas and California, has swept the world and spawned a furiously competitive global market. Annual sales of the integrated circuits exceed $12 billion, and most of the $100 billion electronics business is dependent on the chip. A quarter-century after its unspectacular coming-out party, the integrated circuit is regularly referred to in the popular press as "the miracle chip." The miracle has become ubiquitous. The average American home contains dozens of integrated circuits; the average American garage has half a dozen more. The same holds true in virtually every developed country. The 5,000 man-made objects presently floating in space are crammed with millions of integrated circuits and would not be up there if they weren't.

The integrated circuit was an enormous success because it solved an enormously important problem—the tyranny of numbers. But the success story was also a matter of timing. The chip came along just when the computer was growing up, and chips turned out to be the perfect tools for the digital math and logic that computers use. "The synergy between a new component and a new application generated an explosive growth for both," Bob Noyce wrote in a retrospective article two decades after The Monolithic Idea was born. "The computer was the ideal market . . . a much larger market than could have been provided by the traditional applications of electronics in communications."

In traditional circuitry, involving discrete components wired together, resistors and capacitors were cheap, but switching components such as vacuum tubes and transistors were relatively expensive. This situation was nicely suited to the manufacture of radios, television sets, and the like; an ordinary table radio of the 1950s used two or three dozen capacitors and resistors but only two or three transistors. With integrated circuits, the traditional economies were reversed. Resistors and capacitors, which use up power and take up a lot of room on a chip, became expensive; transistors were compact, simple to make, and cheap. That situation is precisely suited for computers and other digital devices, which need large numbers of switches—transistors—and small quantities of other components.

The innards of computers, calculators, video games, etc., consist of chips containing long chains of transistors that switch back and forth to manipulate information. Like a light switch on the wall, these transistors can be either on or off; there's nothing in between. Since there are only two possible conditions, a computer has to reduce every job, every decision, every computation to the simplest possible terms: on or off, yes or no, stop or go, one or zero. Humans can do the same thing, of course. We do it on Easter morning when the kids look for hidden eggs and their parents provide only two clues: "You're hot" or "You're cold." Eventually, most of the eggs are found, but the process is so tedious that even the kids get fed up with it pretty quickly. Computers, in contrast, use this tedious system all day, every day. They have to. They can't handle anything else.

For all the mystique of "electronic brains" and "artificial intelligence," digital devices are actually mindless dullards that rely on computational techniques mankind abandoned in Neanderthal days. Digital problem-solving involves simple math—far simpler than the stuff humans learn in grade school. A computer approaches every problem like a child counting on his fingers; but the computer counts as if it had only one finger. (The word "digital" comes from the Latin *digitus,* meaning "a finger.") The real miracle of the "miracle chip" is that people have devised ways to manipulate this minimal skill so that machines can carry out complex operations.

Although digital devices can recognize only two numbers—1 and 0—they can arrange those two to represent any number in the universe. Large computers have, for example, computed the value of pi to 10,013,395 decimal places—a number so long it would take about four pages of this book just to print it. But the only numbers the computers had to work with were 0 and 1.

To understand how this works, forget numbers for a moment and think about words—specifically, the word "cat." Anyone familiar with languages knows that there is nothing inherent in the purring, four-footed feline species that requires it to be represented by the three Roman letters *C, A,* and *T*. You can spell the thing *gato* or *chatte* and it is still the same animal. For that matter, there's no reason why a cat has to be represented by letters chosen from our 26-letter Roman alphabet. In the story *On Beyond Zebra,* Dr. Seuss tells of a boy who decided that the Roman alphabet was incomplete, so he went on past *Z* and invented 19 new letters, including one, "thnad," that is used to spell the name of a typically Seussian breed of cat. An alphabet can have as many letters in as many different shapes as its users find convenient. When Samuel F. B. Morse perfected his telegraph, he found that only two letters—dot and dash—could be conveniently sent through the wires. With those two, he invented his own alphabet; telegraphers used to practice their skill by tapping out the Morse Code version of cat •–• •– –. The Japanese *kana* alphabet has 47 different letters; in that system a cat is represented by the symbols ねこ, but it means the same thing as our word "cat." The Chinese alphabet includes about 25,000 different characters, including one that stands for cat (猫). The 26-letter alphabet we use is just a selection of symbols that Western culture has grown used to. The word "cat" is just a convenient combination of those symbols that English speakers have settled on to represent a cat.

The same principles apply to numbers. There's nothing inherent in the number 206, just to choose a number at random, that requires it to be represented by the symbols 2, 0, and 6. We just happen to use that representation because of the way our number system works. Our system uses ten different symbols, or digits, for numbers:

0 1 2 3 4 5 6 7 8 9

To represent numbers larger than nine, we add a second column and run through the list again:

10 11 12

Since the turning point—the new column—comes at the number 10, our system is called the "base-10," or decimal, number system. The decimal system is the most familiar counting system in the world today, but it is by no means the only one available. It would be quite simple to go On Beyond Nine and invent new symbols; a duodecimal, or base-12, number system, could look like this:

0 1 2 3 4 5 6 7 8 9 & # 10

With twelve different symbols, the turning point—the new column—comes at the number 12.

The great breakthrough that permitted man to count far beyond 10 with just ten different symbols was the invention of this turning point—a concept the mathematicians call "positional notation." Positional notation means that each digit in a number has a particular value based on its position. In a decimal number, the first (farthest right) digit represents 1's, the next digit 10's, the next 100's, and so on. The number 206 stands for six 1's, no 10's, and two 100's:

100	10	1
×2	×0	×6

Add it all up:

200 + 0 + 6

and you get 206. This number, incidentally, demonstrates why mathematicians consider the invention of a symbol that represents nothing (i.e., the number 0) to have been a revolutionary event in man's intellectual history. Without zero, there would be no positional notation, because there would be no difference between 26 and 206 and 2,000,006. The Romans, for all their other achievements, never hit on the idea of zero and thus were stuck with a cumbersome system of *M*'s, *C*'s, *X*'s, and *I*'s which made higher math just about impossible.

With positional notation, we can use any number of different symbols to count with. We could devise a numerical alphabet with 26 different digits, or 206 different digits, or 2,006 different digits. The base-10 system we use is just a convenient method that people have settled on to represent all numbers.

If you want to know why modern man has settled on a base-10 number system, just spread your hands and count the digits. All creatures develop a number system based on their basic counting equipment; for us, that means our ten fingers. The Mayans, who went around barefoot, used a base-20 ("vigesimal") number system; their calendars employ 20 different digits. The ancient Babylonians, who counted on their two arms as well as their ten fingers, devised a base-12 number system that still lives today in the methods we use to tell time and buy eggs. Someday a diligent grad student doing interdisciplinary work in mathematics and history of film may produce a dissertation demonstrating that the residents of E.T.'s planet use an octal number system; the movie showed plainly that E.T. has eight fingers. For Earth-bound humans, however, the handy counting system is base-ten.

A computer's basic counting equipment is simpler. It is an electronic switch—a transistor—that can be either on or off. Each of these conditions represents one digit; ON represents 1 and OFF represents 0. Such a number system, using only the two digits 1 and 0, is called the "base-2," or binary, system. Just as people can count to any number, no matter how high, with just ten digits, a computer can count to any number with just two. Like people, computers do this through positional notation. Counting in binary starts out just like decimal:

0 1

But in a binary number, the turning point—the new column—comes at the number 2. In binary the two-digit number 10 stands for one two and no 1's—that is, the quantity 2. Binary 11 means one 2 plus one 1 —that is, 3. Another column must be added to write 4, another for 8:

0 1 10 11 100 101 110 111 1000

Things go on this way until we get to the number 1111. Reading from the right (generally the easiest way to read a binary number), 1111 stands for one 1 plus one 2 plus one 4 plus one 8, or 15. Having come to 1111, the system is out of digits again, so another column is added: the number 10000 is the binary version of 16.

In a binary number, in other words, the first (far right) column represents 1's, the second column 2's, the third 4's, the fourth 8's, the fifth 16's, and so on, as long as necessary. The binary number 11001110, just to choose a number at random, represents (from the right), no 1's, one 2, one 4, one 8, no 16's, no 32's, one 64, and one 128:

	128	64	32	16	8	4	2	1	
	×1	×1	×0	×0	×1	×1	×1	×0	
Add it all up									
	128 +	64 +	0 +	0 +	8 +	4 +	2 +	0	= 206

and it turns out that 11001110 is precisely the same number that is represented by 206 in the decimal system. The quantity hasn't changed; the only thing different is the alphabet, or number system, used to represent it.

Except for someone who has lost nine fingers in an accident, a number system based solely on 1 and 0 is not particularly useful for humans. But the binary system is perfect for digital machines, not only because all numbers can be represented by chains of ON and OFF electronic switches, but also because binary arithmetic is the simplest possible mathematical system. Humans learning basic arithmetic in the elementary grades find that our decimal system requires them to memorize hundreds of "math facts"—facts like $2+2=4$, $17-9=8$, $8\times7=56$. If the school kids were learning the binary system, things would be infinitely simpler. In binary arithmetic, there are only three math facts: $0+0=0$, $1+0=1$, and $1+1=10$ (10, of course, is the binary version of decimal 2).

This extreme simplicity is a boon to computer designers. It is possible, but cumbersome, to construct extensive mazes of transistors inside a machine to perform binary subtraction, multiplication, and division. Thanks to some convenient mathematical gimmickry, however, digital machines based on the binary system can carry out any numerical operation using only addition, which involves fairly simple circuitry. With an ancient trick called "ones-complement subtraction," computers can solve subtraction problems by adding. Multiplication is performed the way humans did it eons ago, before they developed the multiplication table—by repeated addition. If you ask your calculator to multiply 4 × 1,000, the machine will put a binary 4 in its adder unit and then proceed to add 999 more 4's, one at a time, to get the answer. Division, similarly, becomes a series of ones-complement subtractions.

The ingenious, indeed breathtaking, insight that binary mathematics was perfectly suited to electronic computers occurred more or less simultaneously on both sides of the Atlantic to a pair of ingenious, indeed breathtaking, visionaries who had scoped out by the late 1940s remarkably accurate forecasts of the development of digital computers over the ensuing half-century. These two cybernetic pioneers were John von Neumann and Alan M. Turing.

Von Neumann was born in Budapest, the son of a wealthy banker, in 1903. He was recognized almost immediately as a prodigious mathematical talent, and spent his youth shuttling from one great university to another: Berlin, Zurich, Budapest, Göttingen, Hamburg. He published his first scholarly monograph at the age of eighteen and thereafter turned out key papers in a wide variety of fields. In 1930 he sailed west with a tide of refugee European scholars to Princeton, where he held a chair at the university but also became one of the first Fellows—along with Albert Einstein—of the Institute for Advanced Study. He made important contributions in pure mathematics but also wrote major works on applications, ranging from chemical engineering and quantum physics to economics and the "Theory of Games," a mathematical construct of his own for winning complex games.

During World War II, Von Neumann was involved with development of nuclear bombs, an engineering task of overwhelming scope that required, among much else, huge numbers of separate mathematical computations. On a train platform in Aberdeen, Maryland, one day in 1944, Von Neumann was pondering how best to scale this mountain of mathematics when, by sheer luck, he ran into a younger mathematician, Herman Goldstine. Chatting on the train, Goldstine told Von Neumann about the new ENIAC computer under way in Philadelphia—a machine that could zip through repetitive computations at unprecedented speed. This chance conversation pulled Von Neumann into the new world of computers, where he immediately began making major contributions.

When the far-sighted Von Neumann looked on ENIAC and the other primitive, severely limited computers of the late 1940s, he could see a future in which computing machines had almost no limits. More and more, toward the end of his life, he began to see parallels between the evolution of computing machines and the evolution of the human mind. His last book, published posthumously in 1958, was titled *The Computer and the Brain.*

Turing, born in London in 1912, was considered a poor student with little academic promise through most of his school career. After twice failing the scholarship exam for Trinity College, Cambridge, he matriculated at King's, another Cambridge college, and took his Ph.D. there in 1935. He became intrigued by the *Entscheidungsproblem,* a deep mathematical quandary posed by the German scholar David Hilbert. While pondering Hilbert's problem, Turing hit upon an extraordinary new idea: that a machine could be designed, or programmed, to perform any mathematical computation a human could carry out as long as there was a clear set of instructions for this machine to follow. This "Turing Machine" became a key influence and inspiration for computer pioneers in Europe and the U.S.—among them Von Neumann, whom Turing met during a stay at Princeton in the mid-thirties.

During the war, Turing joined the team of mathematicians who gave the Allies an invaluable step up by cracking the Germans' "Enigma" military code. The work involved reading pages and pages of sheer gobbledygook, looking for repetitive patterns of letters that would reveal, under ingenious mathematical manipulation, the inner workings of the German cipher machines. To carry out the calculations, the code breakers developed simple mathematical machines of their own—real-life variations on the abstract Turing Machine. After the war, Turing worked on the first generation of British computers. At the age of forty-one, shortly after being tried and convicted for homosexual conduct ("ACCUSED HAD POWERFUL BRAIN," a London tabloid reported), he died from eating an apple tainted with cyanide he was using for an experiment.

Turing had gone even further than Von Neumann in suggesting that electronic "brains" could eventually match those of their human builders. "One day ladies will take their computers for walks in the park," he said, "and tell each other, 'My little computer said such a funny thing this morning.' " In his most famous paper, "Can a Machine Think?" published in 1951, he predicted that computers would be carrying on "human" conversations with men and with other machines "in about fifty years' time." The filmmaker Stanley Kubrick read that monograph, did the addition, and went to work on a movie about the year 2001.

Von Neumann, in a report to the U.S. Army Ordnance Department, and Turing, in a report for England's National Physical Laboratory, set forth their notions of the general architecture of an electronic computer. Both agreed that the device would have to carry out four basic functions: "input," to take in data and instructions; "memory" (in England, "store"), to keep track of the data; "processing," to do the actual computing; and "output," to report the answer back to the human user.

And both concluded that the logical way to handle data was in the form of binary numbers. Since computer calculations would be performed by switches flipping from on to off and back again, Turing wrote, it was natural enough to assign the value 1 to on and 0 to off and handle all mathematics with only those two digits. "We feel strongly in favor of the binary system for our device," the Von Neumann report agreed. The computer "is naturally adapted to the binary system since we . . . are content to distinguish [just] two states," he wrote. "The main virtue of the binary system . . . is, however, the greater simplicity and speed with which the elementary operations can be performed."

For all the virtues of binary, though, there was a problem—something the mathematicians called the "conversion problem." This was a euphemism for the fact that few humans understand the binary system and thus would find it difficult to convert a computer's answers into a form intelligible to people.

Two solutions were proposed. One was that the human race should drop its decimal system and learn something closer to binary. Under this arrangement, preschoolers watching Sesame Street would be indoctrinated with rhymes like this:

1 10
Buckle my shoe

11 100
Shut the door,

101 110
Pick up sticks

111 1000
Lay them straight

1001 1010
The big fat hen.

Such a sharp change in human habits was obviated by a more practicable idea set forth by both Turing and Von Neumann. "The one disadvantage of the binary system from the human point of view," Von Neumann's report noted, "is the conversion problem. Since, however, it is completely known how to convert numbers from one base to another . . . there is no reason why the computer itself cannot carry out this conversion."

This suggestion was quickly adopted, and ever since all digital devices have included a piece of circuitry called a "decoder," which translates decimal numbers into their binary equivalents. When you punch the keys to put the number 206 into your calculator, the decoder sends out electronic pulses to a chain of eight transistors so that the transistors line up this way:

ON	ON	OFF	OFF	ON	ON	ON	OFF
1	1	0	0	1	1	1	0

Thus transformed to binary format—"11001110"—the number 206 becomes comprehensible to a digital machine.

It was human genius on the part of Von Neumann, Turing, and others like them that figured out how to use binary numbers and binary math to turn an inert chain of electronic switches into a powerful computational tool. But the computer pioneers did not stop there. They also designed a complete system of logic that permits machines to make decisions and comparisons and thus work through complex patterns, or "programs," for manipulating words and numbers. The beauty of this logic system is that it, too, is binary; it can be implemented by integrated circuits full of transistors that do nothing but switch on and off.

Which modern high-tech genius developed this binary logic? None. The logical methods that all digital devices use today were worked out about 100 years B.C. (before computers) by a British mathematician named George Boole.

Boole was born in Lincolnshire in 1815, the son of a cobbler who was always pressed for money. The family's lowly status dictated that the boy would enter some manual trade; he was sent to a vocational school that did not even attempt to teach Latin, the *sine qua non* of a professional future for any English lad of that day. Undaunted, George taught himself Latin and Greek after school. This came in handy in 1831, when the sixteen-year-old boy was forced to leave school and help support the family. He took a job as an assistant teacher but continued to educate himself. In his seventeenth year Boole had two experiences that changed his life. He read Newton's *Principia* and transferred his attention from classical languages to math. Shortly afterward, while walking alone through an open field, Boole was suddenly struck with a "flash of psychological insight" that convinced him that all human mental processes could be formulated in straightforward mathematical terms.

It would be pleasant to report that Boole then and there dedicated his life to the explication of this great concept. Unfortunately, things were not that easy. The family was now dependent on George for support, and his job left insufficient time for complex mathematical work. His knowledge of Latin and Greek qualified him for the clergy, and he decided to train for ordination. Gradually, though, it became clear that Boole was too much a freethinker for such a career; unlike the Anglican Church, he doubted the literal truth of the Bible and believed in religious tolerance. All his life, in fact, he was suspicious of clergymen and their efforts at indoctrination. On his deathbed, according to a biographer, William Kneale, Boole requested "that his children not be allowed to fall into the hands of those who were commonly thought religious."

Committed to teaching, Boole opened his own school at Lincolnshire when he was twenty and now found some time for mathematical work. The editor of a new journal was willing to publish Boole's papers, despite the author's lack of formal training. One of them caught the eye of the mathematician Augustus De Morgan, who helped Boole obtain a chair at Queen's College in Ireland. At last Boole had a secure income and the time to work out his grand mathematical synthesis of human thought. In 1854, after a decade of intense work, he published his masterwork, *The Laws of Thought, on Which Are Founded the Mathematical Theories of Logic and Probabilities.* "The design of the following treatise," the book begins, "is to investigate the fundamental laws of those operations of the mind by which reasoning is performed; to give expression to them in the symbolic language of a Calculus; and upon this foundation to establish the science of Logic. . . ." Completely new and somewhat obscure even to the expert, it had little initial impact. Today it is recognized as a milestone that did indeed establish the new science of symbolic logic.

In Ireland, Boole married Mary Everest, niece of Sir George Everest, the geographer who surveyed the high mountains of Nepal and left his name to the highest of all. In addition to his work on logic, Boole published two widely used textbooks and countless monographs. He found time for poetry, including a difficult lyric entitled "Sonnet to the Number Three." He was a trustee of the Female Penitents' Home and an officer of the Early Closing Association, which strove to reduce the workday to ten hours. A photograph shows him to be an intense, thoughtful professor with a square face, dark hair, and penetrating eyes. For all his achievements in higher math, he never shirked his duties as a teacher. In November of 1864 he walked two miles through a cold rain to meet a class and proceeded with the lecture in his sodden clothes. From this he contracted pneumonia and died. He left behind one last manuscript, so singular and so arcane that the experts at the Royal Society could not decipher it. "No mere mathematician can understand it," his widow observed, "and no theologian cares to try."

Since Boolean logic—also known as Boolean algebra, because Boole expressed logical concepts in algebraic terms—is now recognized as something important, the academicians have draped it in a formidable veil of complicated jargon, symbols, and formulas. At the core, though, the Laws of Thought that Boole described in mathematical terms are the stuff of everyday life. Boole examined everyday mental processes in terms of the simple connective tissue of language: and, or, not.

You wake up from a sound sleep. Can you roll over and sleep some more, or do you have to get up and go to work? To decide, you carry out a fundamental Boolean operation. If your clock says Yes, it's after 8:00, and your calendar says Yes, it's a weekday, then Yes, you get up for work. If either of these conditions is a No, however, you can stay in bed. This decision is known today as a Boolean AND operation. The result is Yes only if condition 1 AND condition 2 are both Yes.

The kitchen sink represents another basic Boolean pattern. If no faucet is on, no water comes out of the spigot. But if either the hot faucet OR the cold faucet is on, OR if both are on, water will flow. This decision, in which the result is Yes if condition 1 OR condition 2 OR both are Yes, is known as a Boolean OR operation.

In essence, Boole demonstrated that all human reasoning could be reduced to a series of yes-or-no decisions. Each decision could therefore be represented in algebraic terms. Sometimes the formulas were as simple as $x+y=z$, and sometimes they were more complex; in *The Laws of Thought* Boole formulates an argument that God exists as "$x(1-y)\ (1-z)+y(1-x)\ (1-z)+z(1-x)\ (1-y)=1$." The most important of Boole's algebraic formulas—the one he describes as the central pillar of his entire yes-or-no structure—is this:

$$x = x^2$$

Anyone young enough to remember high school algebra will see that this equation holds true for two, and only two, numbers: 0 and 1. In other words, Boole's organization of all human decisions into yes-or-no terms turned out to be a binary system. The self-taught Victorian scholar had developed a decision-making methodology that would prove just right for digital machines.

Until digital machines came along, however, Boole's algebra was largely ignored, except by a few of his fellow logicians. One of Boole's most avid followers was the Oxford mathematician Charles Lutwidge Dodgson, who wrote a series of academic works on symbolic logic and who, under his pen name, Lewis Carroll, sprinkled his "Alice" books with allusions to Boole's ideas. Many of the people Alice meets beyond the looking glass see their world in basic Boolean terms—yes-or-no, true-or-false, does-or-doesn't:

> "You are sad," the knight said in an anxious tone. "Let me sing you a song to comfort you."
>
> "It is very long?" Alice asked, for she had heard a good deal of poetry that day.
>
> "It's long," said the Knight, "but it's very, *very* beautiful. Everybody that hears me sing it—either it brings the *tears* into their eyes, or else—"
>
> "Or else what? . . ."
>
> "Or else it doesn't, you know."

Bertrand Russell was another admirer. In *Principia Mathematica,* the Promethean effort to set down once and for all the fundamental logical basis of all mathematics, Russell and Alfred North Whitehead carried Boole's original concept to a climactic conclusion. Unstintingly meticulous (it takes the authors one and a half volumes to arrive at their proof that $1+1=2$) and inaccessible to all but a small coterie of experts, the Russell-Whitehead treatise seemed to offer further proof, if any were needed, that Boole's curious combination of logic and algebra was an intellectual abstraction devoid of practical use. This was hardly what Boole had intended: "The abstract doctrines of science," he wrote in *The Laws of Thought,* "should minister to more than intellectual gratification." Fifty years after it appeared, though, Boole's great work was considered strictly an academic exercise.

The narrative now shifts ahead to 1937 and across the Atlantic to Cambridge, Massachusetts, where groups of engineers and mathematicians were struggling to design the first primitive versions of a digital computer. An M.I.T. engineer, Vannevar Bush, had designed an electrical calculating machine that used decimal numbers; it was built of rods, shafts, and gears arranged so that a gear would turn one-tenth of a full rotation (36 degrees) to represent the number 1, two-tenths (72 degrees) for 2, and so on. This device, although revolutionary for its day, tended to be imprecise; if the gear happened to turn 48 degrees, or 55 degrees, what number was represented? And so attention shifted down the street to Harvard, where another engineer, Howard Aiken, was thinking—as Von Neumann and Turing had been—about a binary machine that would use simple electrical switches. On the binary computer, precision was not a problem—the switches were either on or off, nothing else—but it was a forbiddingly complicated task to design the proper combinations of switches to carry out binary arithmetic.

An M.I.T. graduate student, Claude E. Shannon, who had been working with Bush, was looking for a thesis topic and decided to take on the important but formidable problem of designing digital switching circuits. In the course of his work, Shannon hit upon a crucial idea.

If society allocated fame and fortune on the basis of intellectual merit, Claude Shannon today would be as rich and famous as any rock idol or football star. Born in the farm community of Gaylord, Michigan, in 1916, he graduated from the University of Michigan in 1936 and went on to take a Ph.D. in electrical engineering at M.I.T. His master's thesis, in 1937, demonstrated how computer circuits should be designed and launched a new academic discipline known as "switching theory." Ten years later, as a researcher at Bell Labs, he got thinking about efficient means of electronic communications (for example, how to send the largest number of telephone conversations through a single wire). He published another seminal paper that launched an even more important new discipline known as "information theory"; today information theory is fundamental not only in electronics and computer science but also in linguistics, sociology, and numerous other fields. A year after that paper, Shannon published a monograph—it was the first on the topic—called "Programming a Computer for Playing Chess." The ideas set forth there are still central to the design of all computer games. Like many mathematicians, Shannon is an avid fan of games and puzzles; among other things, he likes to work out "pangrams"—sentences that contain every letter of the alphabet. His *pièce de résistance* in this field is a sentence that uses each letter only once: "Squdgy fez, blank jimp crwth vox!" Each of the words is in the (unabridged) dictionary; a letter to Shannon at M.I.T., where he is now emeritus professor of electrical engineering, will bring a complete explanation.

In 1937, when Shannon tackled the problem of binary circuit design, digital computers used magnetic switches called "relays." A relay looks like a mousetrap with an electromagnet on one end. When electricity flows to the magnet, it attracts the metal bar of the mousetrap, which flips over and thus turns the switch on. As soon as the current is cut off, the magnetic attraction stops and the metal bar springs back, turning the relay off. The problem was how to design arrays of these relays so they would switch on and off in the proper order to add binary numbers.

Stewing over this question, Shannon happened upon a text on Boolean logic—and something clicked. Boole's equations for AND operations, OR operations, and other logical functions reduced decision making to a set of dualities—yes or no, 0 or 1, true or false. Shannon recognized that these pairs could be represented just as well by the switching duality: on or off. In short, the forbidding job of designing binary logic circuits had already been done—by George Boole. Boole's carefully worked-out equations could serve as road maps for wiring together electric switches to carry out logical operations. Accordingly, Shannon wrote, "It is possible to perform complex mathematical operations by means of relay circuits. Numbers may be represented by the positions of relays and stepping switches, and interconnections between sets of relays can be made to represent various mathematical operations." At the end of his paper, Shannon showed how a series of relays arranged to carry out AND and OR operations could be wired to add two binary numbers.

In addition to mathematical operations, Shannon demonstrated, Boolean circuits could also be wired to make comparisons—is number x equal to number y?—and to follow simple directions of the "If A, then B" category. "In fact, any operation that can be completely described in a finite number of steps using the words 'if,' 'or,' 'and,' etc.," Shannon wrote, "can be done automatically with relays." With this ability to make decisions—to proceed in different ways depending on the results of its calculations—the machine could be programmed to carry out complicated computations without constant direction from the human operator.

The techniques set forth in Shannon's thesis have been universally adopted for digital machines. In modern computers transistors embedded in integrated circuits have replaced magnetic relays, but the principles of binary switching remain the same. A transistor built into integrated circuits is essentially the same silicon sandwich developed by Shockley's team: it consists of a thin layer of P-type silicon sandwiched between two slightly thicker layers of N-type silicon. The device is hooked up so that current—a surge of electrons—will run from one N-type layer, through the middle, and out the other N-type layer. This current flow is switched on and off by signal pulses flowing to the middle layer. If a pulse is sent to the center layer, current will flow from end to end; the transistor is on. But if the center receives no pulse, it blocks current flow from N to N; then the switch is off.

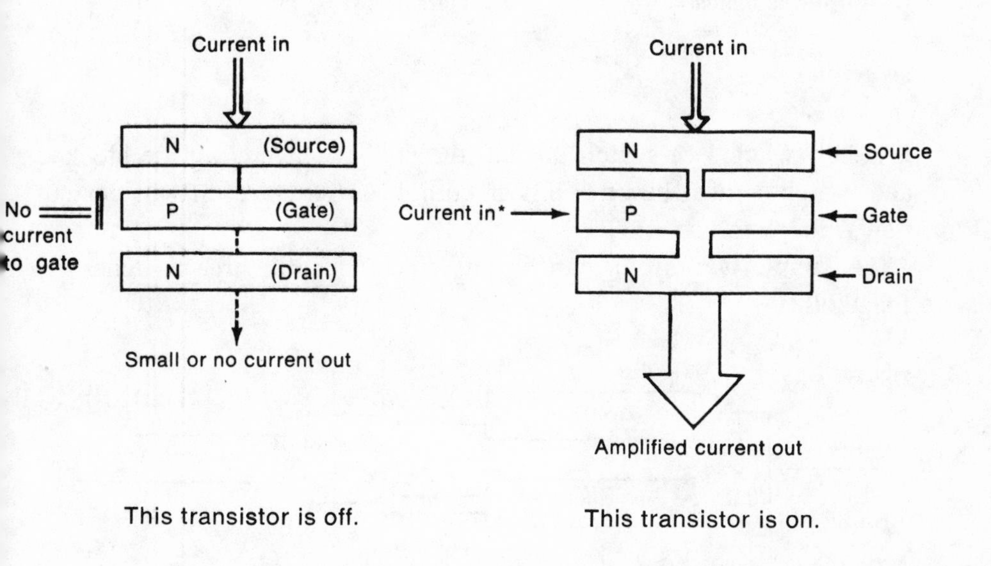

This transistor is off. This transistor is on.

* The greater the current flowing into the gate, the greater the flow from source to drain.

A computer's circuitry is a chain of transistors, one after another. The circuit is analogous to a long irrigation pipe with a series of faucets built into it to control the flow of water. If faucet 1 is open, water can flow along to faucet 2; if that one is open, water can flow on to faucet 3. By opening the right combination of faucets at the right time, a farmer can direct water to any point in his field. In a computer's electronic pipeline, each transistor acts as a faucet; if transistor 1 is on, current can flow on through to transistor 2, and so on. By turning on the right combination of transistors at the right time, computer designers can direct the flow of current to any point in the circuit.

To set up the right combinations, computer builders rely mainly on three basic Boolean circuits. The simplest AND circuit, in accordance with Boole's AND operation, consists of three transistors lined up so that switch 3 is on only if both switch 1 AND switch 2 are on. The arrangement looks like this:

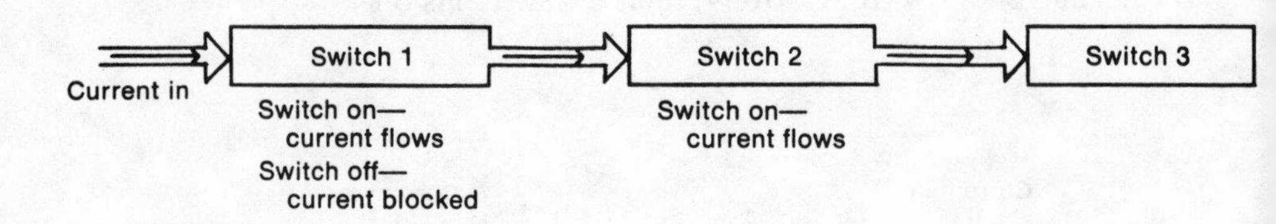

If either switch 1 or switch 2 is off, the flow of current will be blocked and switch 3 must be off. Only if both 1 AND 2 are on will current flow through to turn on 3.

A simple OR circuit can be implemented with three switches arranged this way:

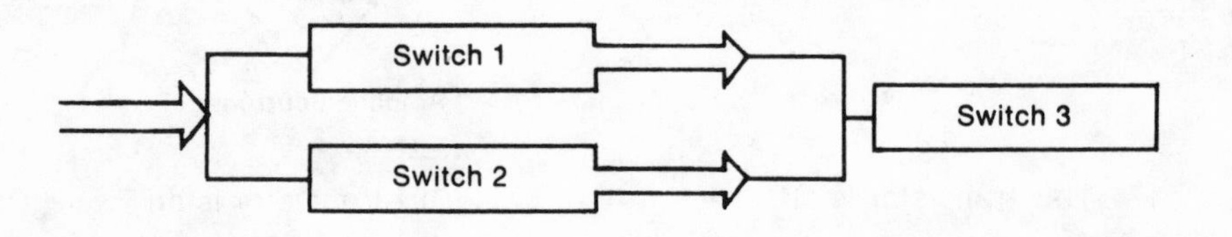

If both switch 1 and switch 2 are off, current flow will be blocked and switch 3 must be off. But in this circuit, current can flow through either switch 1 or switch 2 to get to switch 3. Thus if either 1 OR 2 is on, OR if both are on, current will flow through the circuit to turn on 3.

The third basic circuit, called a NOT circuit, can be wired from two switches arranged so that the second is NOT in the same state as the first. If switch 1 is on, 2 is off; if 1 turns off, 2 switches on.

Because these switch arrangements serve either to block current or to let it pass, they are commonly called "gates." Logical and mathematical operations are carried out by sending current through a maze of different gates. A basic addition circuit, in its simplest form, can be implemented with a dozen AND gates, a half-dozen OR gates, and three NOT gates. Pulses representing the binary numbers to be added are sent into the circuit. Each of these pulses turns selected transistors on or off in just the right combination so that the pulses coming out at the end of the circuit will represent, in binary, the sum of the two numbers that went in.

Mathematicians like to use the term "elegant" to describe a simple solution to a complex problem. The Boolean logic that adds two binary digits is the height of elegance. In the early computers, though, the electronics of this elegant arrangement were extremely cumbersome. The simple addition circuit just described, with its twenty-one separate "gates," requires about fifty transistors, a dozen or so other components, and a labyrinthine spaghetti of connecting wires. And this circuit can add only two binary digits. If the numbers being added require, say, eight binary digits each—like the decimal number 206, which is "11001110" in binary—the problem would require eight separate passes through the addition circuit, plus a few dozen more transistors to store the result of each pass.

This is why computers were so vulnerable to the tyranny of numbers. The use of binary numbers and binary logic provided a precise computational system that was perfectly suited to electronic devices. But this perfect fit came at a price. The price was complexity. Digital devices are nothing more than switches turning on and turning off, but large numbers of switches must turn on and off large numbers of times to perform even simple operations. When electronic circuits had to be hand-wired together from individual components, these large numbers took an enormous toll in size, in speed, in cost, in power consumption, in difficulty of design.

Despite its inauspicious debut, accordingly, The Monolithic Idea, conceived just as the digital computer was growing up, was destined to be a spectacular success. With integrated circuitry, the neat patterns of Boolean logic could be mapped directly onto the surface of a silicon chip; an entire addition circuit would now take up less space and consume less power than a single transistor did in the days of discrete components. With the advent of the chip, the digital computer had finally become as elegant in practice as it was on paper.

7

Blasting Off

THE FIRST INTEGRATED CIRCUITS proved so hard to produce that nearly two years passed after the chip's public debut before the new device was available for sale. Fairchild was first off the block; its catalogue for spring of 1961 trumpeted a new line of six different monolithic circuits which it called "Micrologic Elements." A few weeks later Texas Instruments entered the fray with a similar series of "solid circuits." As the two companies rather stridently pointed out, the new circuit-on-a-chip was smaller, lighter, faster, more power efficient, and more reliable than any conventional circuit wired together from discrete parts.

It was also more expensive. A "micrologic" logic gate circuit, containing three or four transistors and another half-dozen diodes and resistors, was initially priced at $120. An equipment manufacturer could wire together a circuit using top-of-the-line transistors for less than that, even after labor costs were figured in. It was as if an automobile company had designed a family station wagon that could go 500 miles per hour—and cost $50,000. Who needed it? "There was the natural reluctance to commit to something completely new," Bob Noyce recalled later. "And added to that you had a price that was basically uneconomical. So at first the traditional electronics customers just weren't buying."

This posed a fairly serious problem. Even more than most industries, electronics firms rely heavily on an economic phenomenon known as the "learning curve." In the early life of a new product—when manufacturers are still learning how to design and produce the device at a reasonable cost—prices are necessarily high. As sales increase, better production techniques are developed, and prices curve sharply downward. The integrated circuit in early 1961 was stalled at the high end of the curve; there was no commercial market to push it down. As Noyce recalls, the chip seemed to be caught in a classic commercial Catch-22. Until the market picked up, the price would remain high; but as long as prices stayed high, the traditional electronic markets weren't interested.

And then, virtually overnight, the President of the United States created a new market.

In May of 1961—a time when, as *The New York Times* noted, "there was a strong catch-the-Russians mood in Washington"—John F. Kennedy went before a joint session of Congress to propose "an extraordinary challenge." "I believe we should go to the moon," the president said. ". . . I believe that this nation should commit itself to achieving the goal, before this decade is out, of landing a man on the moon and returning him safely to earth. No single space project in this period will be more impressive to mankind. . . . And none will be so difficult or expensive to accomplish."

A 480,000-mile round trip to the moon was indeed a challenge of extraordinary dimensions for a nation whose greatest space achievement so far had been Alan Shepard's 15-minute, 302-mile suborbital flight in the spacecraft *Freedom 7*. (John Glenn, the third American in space, did not make his three-orbit trip until nine months after Kennedy's speech.) A successful lunar voyage would require major advances in rocketry, metallurgy, communications, and other fields. Among the most difficult problems were those the space experts called "G & N"—i.e., Guidance and Navigation.

The trick of steering a fast-moving spaceship from a fast-moving planet through two different atmospheres and two different gravitational fields to a precise landing on a fast-moving satellite would require an endless series of instantly updated calculations—the kind of work only a computer could do. But a computer on a spacecraft would have to be smaller, lighter, faster, more power-efficient, and more reliable than any computer in existence. In short, somebody needed that 500-mph station wagon. And with the prestige of the nation at stake, high prices were no problem. "The space program badly needed the things that an integrated circuit could provide," Kilby said later. "They needed it so badly they were willing to pay two times or three times the price of a standard circuit to get it."

The G & N system was so crucial to the moon shot that the assignment for its development was the first major prime contract awarded after Kennedy's speech. There was no question that the computer would have to be built from integrated circuits, and Fairchild quickly started receiving large orders for its Micrologic chips. By the time the Eagle had landed at Tranquility Base on July 20, 1969—meeting the late president's challenge with five months to spare—the Apollo program had purchased more than a million integrated circuits.

Today, when the American semiconductor industry is facing an all-out battle with Japanese competitors, U.S. electronics companies frequently complain that Japanese firms have an unfair advantage because much of their development funds were provided by the government in Tokyo. On this point, the American manufacturers live in a glass house. The government in Washington—specifically, the National Aeronautics and Space Administration and the Defense Department—played a crucial role in the development of the American semiconductor industry. The Apollo project was the most glamorous early application of the chip, but there were numerous other rocket and weapon programs that provided research funds and, more important, large markets when the chip was still too expensive to compete against traditional circuits in civilian applications. A study published in 1977 reported that the government provided just under half of all the research and development money spent by the U.S. electronics industry in the first sixteen years of the chip's existence. Government sales constituted 100 percent of the market for integrated circuits until 1964, and the federal government remained the largest buyer of chips for several years after that.

The military had started funding research on new types of electric circuits in the early 1950s, when the tyranny of numbers first emerged. The problems inherent in complex circuits containing large numbers of individual components were particularly severe in defense applications. Such circuits tended to be big and heavy, but the services needed equipment that was light and portable. "The general rule of thumb in a missile was that one extra pound of payload cost $100,000 worth of extra fuel," Noyce recalls. "The shipping cost of sending up a 50-pound computer was too high even for the Pentagon." Further, space-age weapons had to be absolutely reliable—a goal that was inordinately difficult to achieve in a circuit with several thousand components and several thousand hand-soldered connections. When the Air Force ordered electronic equipment for the Minuteman 1, the first modern Inter-Continental Ballistic Missile, specifications called for every single component—not just every radio but every transistor and every resistor in every radio—to have its own individual progress chart on which production, installation, checking, and rechecking could be recorded. Testing, retesting, and re-retesting more than doubled the cost of each electronic part.

In classic fashion, the three military services went off in three different directions in the search for a solution. The Navy focused on a "thin-film" circuit in which some components could be "printed" on a ceramic base, somewhat reducing the cost and size of the circuit; Jack Kilby worked on this idea for a while during his years in Milwaukee at Centralab. The Army's line of attack centered around the "micro-module" idea—the Lego-block system in which different components could be snapped together to make any sort of circuit. Kilby worked on that one for a few days when he first arrived at Texas Instruments.

The Air Force, whose growing fleet of missiles posed the most acute need for small but reliable electronics, came up with the most drastic strategy of all. It decided to jettison anything having to do with conventional circuits or conventional components and start over. The Air Force program was called "molecular electronics" because the scientists thought they could find something in the basic structure of the molecule that would serve the function of traditional resistors, diodes, etc. Bob Noyce brushed up against Molecular Electronics early in his career. "The idea of it was, well, you lay down a layer of this and a layer of that and maybe it will serve some function," Noyce said later. "It was absolutely the wrong way to solve anything. It wasn't built up from understandable elements. It didn't start with fundamentals because they were rejecting all the fundamentals. It was pretty clearly destined for failure." The Air Force wasn't listening. With strong lobbying from the generals, Molecular Electronics won the official bureaucratic seal of approval—a line item of its own in the federal budget. Congress eventually appropriated some $5 million in research funds. Nothing came of the idea.

Each service, naturally, was eager to see its own approach prevail. All three services, consequently, were somewhat taken aback when they learned, in the fall of 1958, that a fellow named Kilby at Texas Instruments had worked up a solution to the numbers problem that was neither Army nor Navy nor Air Force.

The military services learned of Kilby's new monolithic circuit as soon as the people at Texas Instruments had tested the first chip and found that it worked. "T.I. had always followed a strategy of getting the Pentagon to help with development projects," Kilby explained later. "So sometime in the fall of 1958 Willis [Adcock] and I started telling the services what we had." The Navy wasn't interested. The Army agreed to provide funding to prove that Kilby's new integrated circuit was "fully compatible" with the Micro-Module. "Well, it wasn't a Micro-Module at all," Kilby recalled. "But that was okay; it gave us some money to work with, and we didn't care what they called it. If they wanted it green, we'd paint it green." Hoping to supplement the modest Army grant, Adcock and Kilby spoke to the Air Force. "They weren't interested," Kilby said later. "Our circuit had the traditional components, resistors and the like, and their approach wasn't going to have any of that traditional jazz." Despite the initial rejection, Adcock wouldn't give up. For months he argued his case, and eventually he found a colonel who was starting to lose faith in the cherished notion of Molecular Electronics. In June of 1959, the Air Force consented to provide just over $1 million for developmental work on the chip. (Years later, the Air Force's public relations wing put out a book on microelectronics: "The development of integrated circuits is, in large part, the story of imaginative and aggressive leadership by the U.S. Air Force.")

Events followed a different course at Fairchild, largely because Bob Noyce had different ideas about Pentagon-funded research. Noyce had worked on some defense research and development projects when he was a young engineer at Philco, and the experience left a sour taste that never went away. It wasn't fair, he thought—it was "almost an insult"—to ask a competent, creative engineer to work under the supervision of an Army officer who had at best a passing familiarity with electronics. The right way for a private concern to carry out research, Noyce felt strongly, was with private money. If this research happened to produce something useful for the military, fine—but Noyce did not want his engineers restricted to military research or bound by the confines of a defense development contract.

And so Fairchild developed The Monolithic Idea into a salable commodity using its own funds. Noyce readily concedes, though, that the company was willing to do so in considerable part because of potential sales to the military market. "The missile program and the space race were heating up," Noyce says. "What that meant was there was a market for advanced devices at uneconomic prices . . . so there was a lot of motivation to produce this thing."

In addition to the Apollo program, several new families of nuclear missiles provided large early markets for integrated-circuit guidance computers. The designers of Minuteman II, the second-generation ICBM, decided in 1962 to switch to the chip. With that decision, which led to $24 million in electronics contracts over the next three years, the integrated circuit took off. Texas Instruments was soon selling 4,000 chips per month to the Minuteman program, and Fairchild, too, landed important Minuteman contracts. Soon thereafter the Navy began buying integrated circuits for its first submarine-launched intercontinental missile, the Polaris. By the mid-sixties, chips were routinely called for in specifications for a large variety of military electronic gear—not only G & N computers but also telemetry encoders, infrared trackers, Loran receivers, avionics instruments, and much more. NASA's IMP satellite, launched late in 1963, was the first space vehicle to use integrated electronics, and thereafter chips became the circuits of choice for satellites and other space endeavors. About 500,000 integrated circuits were sold in 1963; sales quadrupled the next year, quadrupled again the year after that, and quadrupled again the year after that.

The burgeoning government sales not only provided profits for the chips makers but also conferred respectability. "From a marketing standpoint, Apollo and the Minuteman were ideal customers," Kilby says. "When they decided that they could use these Solid Circuits, that had quite an impact on a lot of people who bought electronic equipment. Both of those projects were recognized as outstanding engineering operations, and if the integrated circuit was good enough for them, well, that meant it was good enough for a lot of other people."

One of the major pastimes among professional economists is an apparently endless debate as to whether military-funded research helps or hurts the civilian economy. As a general matter, there seem to be enough arguments on both sides to keep the debaters fruitfully occupied for years to come. In the specific case of the integrated circuit, however, there is no doubt that the Pentagon's money produced real benefits for the civilian electronics business—and for civilian consumers. Unlike armored personnel carriers or nuclear cannon or zero-gravity food tubes, the electronic logic gates, radios, etc., that space and military programs use are fairly easily converted to earthbound civilian applications. The first chip sold for the commercial market—used in a Zenith hearing aid that went on sale in 1964—was the same integrated amplifier circuit used in the IMP satellite. For the Minuteman II missile, Texas Instruments had to design and produce twenty-two fairly standard types of circuits in integrated form; every one of those chips was readily adaptable to civilian computers, radio transmitters, and the like. A large number of the most familiar products of the microelectronic revolution, from the busy businessman's pocket beeper to the Action News Minicam ("film at eleven"), resulted directly from space and military development contracts.

The government's willingness to buy chips in quantity at premium prices provided the money the semiconductor firms needed to hone their skills in designing and producing monolithic circuits. With their earnings from defense and space sales, Fairchild, Texas Instruments, and a rapidly growing list of other companies developed elaborate manufacturing facilities and precise new techniques for making chips. As experience taught ways to solve the most common production problems, the cost of making a chip began to fall. By 1964 the initial manufacturing base was in place, and the integrated circuit started flying down the learning curve with the speed of a lunar rocket in reentry. In 1963 the price of an average chip was about $32. A year later the average price was $18.50, a year after that $8.33. By 1971, the tenth anniversary of the chip's arrival in the marketplace, the average price was $1.27.

While prices were falling, capability soared. Year after year, buyers of integrated circuits got more product for less money. Manufacturers learned how to cram more and more components onto a single chip. This achievement was partly a matter of design; complex circuits had to be laid out on the tiny flake of silicon so that each individual component could perform its function without interfering with any of the other components squeezed alongside. The chief technical obstacle to high-density chips, though, was production yield. The more components printed onto a chip, the greater the chance that one of those components would have a defect. One defective transistor could render the entire integrated circuit worthless. "A single speck of dust is huge compared to the components in a high-density circuit," Bob Noyce has said. "One dust particle will easily kill a whole circuit. So you've got to produce the thing in a room that is absolutely free of dust. You've got to build in thousands of [connecting] leads that are finer than a human hair, and every one has to be free of any defect. Well, how do you build a room that's free of dust? And how do you print a lead that is essentially perfect? We had to learn over time how to do things like that."

As the industry learned how, it found itself in a delightful position. A chip containing 10,000 components required no more silicon and not much more labor than one with only 5,000 components; it was as if a publisher had found a way to turn out two books using no more paper or ink than it had previously used for one. And the semiconductor industry found ways to double capacity year after year. Noyce's friend and colleague, Gordon Moore, was asked in 1964—when the most advanced chips contained about 60 components—to predict how far the industry would advance in the next decade. "I did it sort of tongue-in-cheek," Moore recalled later. "I just noticed that the number of transistors on a chip had doubled for each of the last three years, so I said that rate would continue." To his dismay, that off-the-cuff prediction was widely quoted and soon came to be known as "Moore's Law." To his astonishment, the law held up, well into the 1970s. "At the time, I had no idea that anybody would expect us to keep doubling [capacity] for ten more years. If you extrapolated out to 1975, that would mean we'd have 65,000 transistors on a single integrated circuit. It just seemed ridiculous." By 1975 the industry was producing a new series of memory chips that contained 65,536 transistors. "It amazes me," Moore said recently. "I still have a tough time believing that we can make these things."

As the engineers measure it, the integrated circuit has passed through four generations in the quarter-century since Kilby and Noyce first got the idea. The first-generation chips—known as "SSI," for "small-scale integration"—contained fewer than 100 components. The years from 1966 to 1969 are called the era of MSI, the medium-scale-integrated chips ranging from 100 to 1,000 components. The early seventies brought on large-scale integration, a term that includes chips with up to 10,000 components. By the early eighties the industry had mastered very-large-scale integration, producing circuits that had 250,000 components neatly arranged on a sliver of silicon no bigger than the word "CHIP" on this page. ULSI—ultra-large-scale integration, with a million components on a chip—is within striking distance. At some point beyond that (it is not clear exactly where the point might be), the march of integration will have to stop. At some point, there just won't be enough electrons on the chip to trigger any more transistors.

With prices in steep decline and capacity in rapid ascent, the semiconductor industry produced the greatest productivity gains in American industrial history. A graph comparing prices and capacity during the first twenty years of the chip's existence makes a nearly perfect X: the price curve angles sharply downward over time, and the capacity curve angles straight up. The $32 integrated circuits available in 1963 contained about 30 electronic components. Ten years later, a chip that cost a dollar or so would contain well over 1,000 components. From the buyer's standpoint, therefore, the cost dropped by 99.9 percent—from about one dollar to about one-tenth of a cent per component—within the decade. Clearly a thousandfold reduction in price in ten years is something special, and consequently it is probably unfair to compare the chip to other industrial products. The temptation is hard to resist, though, and the comparison is frequently made. In a typical version, Gordon Moore suggested what would have happened if the automobile industry had matched the semiconductor business for productivity. "We would cruise comfortably in our cars at 100,000 mph, getting 50,000 miles per gallon of gasoline," Moore said. "We would find it cheaper to throw away our Rolls-Royce and replace it than to park it downtown for the evening. . . . We could pass it down through several generations without any requirement for repair."

"Progress has been astonishing, even to those of us who have been intimately engaged in the evolving technology," Bob Noyce wrote.

> An individual integrated circuit on a chip perhaps a quarter of an inch square now can embrace more electronic elements than the most complex piece of electronic equipment that could be built in 1950. Today's microcomputer, at a cost of $300, has more computing capacity than the first large electronic computer, ENIAC. It is 20 times faster, has a larger memory, is thousands of times more reliable, consumes the power of a light bulb rather than that of a locomotive, occupies 1/30,000 the volume and costs 1/10,000 as much. It is available by mail order or at your local hobby shop.

Noyce wrote that in 1977. Naturally, the passage is seriously out of date now. *Today's* microcomputer, at a cost as low as $30, has considerably more speed, more memory, and more computing capacity than the $300 device that represented the state of the art when Noyce made the comparison. It is available at J. C. Penney's or your local drugstore.

The dramatic increase in the capacity of a chip also improved circuit performance. Higher-density chips meant less space between components. Less space meant less travel time for signal pulses running from one component to the next, so a smaller circuit was a faster circuit. Similarly, a smaller circuit required less power. Thus the chip makers were offering lower prices, higher capacity, and better performance year after year. In the second half of the 1960s that achievement caught the attention of commercial markets, such as producers of computers and industrial equipment, which had spurned the chip when it first appeared in 1961. By the early 1970s the government was no longer the leading consumer of chips; that position had been taken over by the computer industry.

In retrospect, it is easy to see that the integrated circuit was perfectly suited to the digital computer, but the point was less than obvious to computer manufacturers when the chip first came on the market. In the days before the mini- and micro-computer, when computers routinely cost hundreds of thousands of dollars, the development of a new model represented an enormous investment on the manufacturer's part.

Lead times were long; a decision made in 1961 would govern the production of machines that came on the market four years later. As a result, computer design was a conservative science. In the early 1960s no computer builder was willing to take a chance on a completely new type of circuit.

Thus when IBM brought out a major new line of computers, the System 360, in 1964, the new machine did not use integrated circuits. Still, the System 360 involved a revolutionary concept—it was a *family* of computers in various sizes and prices that shared the same instruction code, or software, and could thus communicate with one another—which rendered all existing machines obsolete. To fight back, competitors had to come up with something new of their own. They turned to integrated logic circuits. Using the chip, Univac, Burroughs, and RCA turned out machines that were as powerful as the IBM system but smaller, faster, and cheaper. Brash upstarts like Digital Equipment and Data General entered the market with a new concept—the fully-integrated device called the "minicomputer." It was about the size of a senior executive's desk and cost less than $100,000 but matched some of IBM's big mainframes in computing power. In 1969 IBM bowed to the inevitable and began using chips for all logic circuitry in its computers. Now the chip makers had a market that would dwarf the space and defense business. By 1970 there were more than two dozen American firms turning out integrated circuits; they sold 300 million chips that year. Two years later 600 million chips were sold.

Logic circuitry, however, represented only one part of the potential computer market. Digital machines need logic gates to manipulate data, but they also need memory units to hold the data. The computers of the 1960s stored data, in the form of binary digits, using an ingeniously simple technique called "magnetic core memory." A core memory looks like a tennis net made of fine wires; wherever two wires cross, a small iron wedding ring—the "core"—is hooked over the intersection. By sending electronic pulses along the right pair of wires, each individual iron core could be magnetized or demagnetized. A magnetized core represented a binary 1; demagnetized, it stood for binary 0. Core memory was fairly bulky: it took about 10 square feet of wire net to store 1,000 bits of information (a "bit" is the computer world's abbreviation for a single *bi*nary dig*it,* a 1 or a 0). But it was reliable, easy to make, and inexpensive. The wires and the iron cores were dirt cheap, and a complete memory unit needed only a handful of transistors to send out the needed pulses. The most expensive thing about a core memory was the labor cost for stringing all those iron rings on the net. This job, done by hand, was eventually farmed out to places like Hong Kong and Mexico, so prices for core memory remained low.

Some farsighted semiconductor engineers could see the possibility of putting memory onto a chip. The integrated circuit, after all, was a perfect medium for storing binary digits. A chip comprised a large array of switches (transistors), and any switch has memory. The light switch on the wall is memory unit; it remembers the last thing you did to it, and stays that way—either on or off—until you change the setting. For various technical reasons, semiconductor memories often used more than one transistor to store each binary digit. In one standard memory design, a block of four transistors was used to store each bit. If a signal pulse turned the block on, it stood for 1; if the block of transistors was off, it stood for 0.

A semiconductor memory chip was ten times smaller than the equivalent core memory unit; since the signal pulses had shorter distances to travel, it was much faster. Through most of the 1960s, however, it was also two or three times more expensive. In 1967, engineers at Fairchild performed the prodigious feat of squeezing 1,024 transistors onto a single integrated circuit. At four transistors per bit, such a circuit could provide storage for 256 bits of information. But a 256-bit memory chip was still more than twice as expensive as a comparable amount of iron core memory. The Fairchild chip was admired in the laboratories but ignored in the market.

A monograph that appeared in the *Proceedings of the Institute of Electrical and Electronic Engineers* in 1968 set forth in discouraging detail the economics of the memory business. Just to approach the price of core memory, it said, the semiconductor people would have to come up with a 1,000-bit memory chip. Semiconductor memory would not actually become cheaper until somebody developed a 4,000-bit chip. Four thousand bits on a single chip? Most of the industry looked at those figures and decided that the wisest course would be to forget memory. A pair of engineers at Fairchild—Bob Noyce and his friend Gordon Moore—looked at the same numbers and decided to give it a try.

When Noyce and Moore left Fairchild in 1968 to start a new company specializing in semiconductor memories, they were gambling that a memory chip would be easier to make in extremely high densities than the traditional logic chips. A logic circuit, with its assorted gates and pathways, requires a variety of components laid out in complex patterns and a byzantine pattern of leads to connect the parts. A memory circuit, in contrast, consisted for the most part of identical transistors lined up in identical blocks, one after another, like blocks of identical tract houses in some suburban Levittown. Connections could be provided by a simple network of parallel leads, just like a neat grid of crisscrossing suburban streets. Each block of transistors, like each house in Levittown, could be assigned a unique address. Consequently, the memory circuit would permit random access—that is, the logic circuits could send data to or extract it from any one of a thousand memory locations without disturbing the other 999.

The new firm that Noyce and Moore founded, Intel Corporation, turned out its first high-density memory circuit late in 1968. It held 1,024 bits of data. Using the standard engineering shorthand for 1,000—the letter K—this Random Access Memory chip was called a "1K RAM." Intel rang up a grand total of $2,672 in sales that first year. By 1973, when the 4K RAM came to market, the firm's sales topped $60 million, and Texas Instruments and several other firms had jumped into the memory business as well.

Like most other fields of human endeavor, the computer world has its own version of Parkinson's Law. It is sometimes stated in pure Parkinsonian terms—"Data expands to fill the memory available to hold it"—and sometimes in plainer language—"There's no such thing as enough memory." For computer buffs, using one bank of memory is like eating one peanut. The computer business turned out to have a voracious appetite for cheap, fast random-access memory, and the semiconductor business geared up to meet the need. Partly because of the commonly used four-transistor-per-bit storage configuration, memory chips tended to grow by factors of four. A 16K RAM came on the market in 1975. The 64K RAM went on sale five years later. The first 256K memory chips were developed in 1983, and the end of the decade should see 1,000K, or "1M," bits of data stored on a single Random Access Memory chip.

In addition to the two types of digital circuits—logic and memory—the late sixties also saw the first significant development of another species of chip, called a "linear" or "analog" integrated circuit. The linear chips replicated the functions of many traditional electronic circuits—timers, radio transmitters, audio amplifiers, and the like. Such applications put the integrated circuit into a number of noncomputer electronic devices. Some were traditional—the first integrated circuit radio receiver went on the market in 1966—and some wholly new—the cardiac pacemaker, a tiny circuit that gives off small electric pulses at precise intervals, was implanted in a human chest for the first time in 1967. By 1970 integrated electronics was replacing traditional circuitry in everything from elevators to Osterizers.

Applications for integrated circuits were multiplying rapidly—but not as rapidly as the semiconductor companies were turning out densely integrated new circuits. "We reached a point where we could produce more complexity than we could use," Gordon Moore said later. Those industry executives who took time to look up from their balance sheets and think about the future could see that supply of electronic circuits was increasing faster than demand. Almost alone among industrial products, moreover, the integrated circuit was a one-time-only sale. There was nothing to break, no moving parts to wear out. Once a chip passed its initial inspection, it would last a lifetime; there was little or no replacement market.

An exam question that pops up now and then in the nation's business schools postulates a situation where somebody invents a common product that will never wear out—a permanent razor blade, for example, or a lifetime light bulb. How should the light bulb industry react? One answer is that the manufacturers should make a quick killing selling the lifetime bulb until all conventional bulbs are replaced—and then go out of business. This answer is not an acceptable one, however, at most American business schools, which preach the importance of unending growth. The right answer is that the industry should use its ingenuity to create new uses for lifetime light bulbs and cater to a continually expanding market.

Faced with an ever-expanding supply of a lifetime product, the semiconductor industry at the end of the sixties picked the right answer. The thing to do was to find new applications and new markets for integrated circuits. Since the chip, up to then, had been sold almost exclusively for government and industrial uses, the obvious new market to shoot for was the largest market of all—the consumer. To maintain its explosive rate of growth, the American semiconductor industry would have to take its revolutionary new product into the American home.

But how? The only chips that most Americans knew much about were made from potatoes; the very word "semiconductor" was completely alien to the general public. How many homes really needed an interplanetary Guidance and Navigation system or a $50,000 computer? Extending the microelectronic revolution down to the average consumer loomed as a formidable problem. To solve it, the industry turned to one of its premier problem-solvers—Jack Kilby.

The Implosion

THE DECISION THAT BROUGHT the chip into virtually every household, and made "chip," in its microelectronic sense, a household word, was a carbon copy of the decision a decade earlier that had done the same things for the chip's immediate ancestor, the transistor. The decision maker in both cases was Patrick Haggerty, the farsighted and plucky chief executive of Texas Instruments. In the early 1950s, when the transistor was starting to become a cheap, reliable mass-production item, Haggerty developed a fascination, almost an obsession, with the notion that microelectronics should play a pervasive role in modern society. He came to believe that microelectronic devices would replace standard circuitry in existing electronic gear; that all tools and appliances controlled by traditional gears, springs, or timers would be fitted with chips; and that the availability of low-cost, low-power, high-reliability miniature components would create entire markets unknown before. "Pervasiveness" occurs like an *idée fixe* in Haggerty's speeches and writings. The conviction that microelectronic devices would gradually pervade every aspect of life was at the root of many of his business decisions.

Haggerty first put the principle into practice in 1954, when Texas Instruments, then a small, regional manufacturer of electronic parts, discovered that it was ahead of everybody else in transistor development. "We knew we were doing pretty well in our semiconductor endeavor," Haggerty recalled twenty-five years later, "[but] we were facing a world that was pretty skeptical. . . . It seemed to me that it was imperative for T.I. to generate some kind of dramatic demonstration that reliable transistors really were available in mass production quantities and at moderate prices, and that T.I. was both ready and able to produce them." In fact, Haggerty had a particularly dramatic demonstration project in mind.

He had called in his engineers and told them to produce something wholly new—a portable radio, completely transistorized, powered by penlight batteries, and small enough to carry in a pocket or install in a dashboard. Haggerty tried to persuade several major radio firms to sell the device. They all demurred, arguing that there was no market for a pocket radio—an accurate assertion, since there had never been a pocket radio.

Eventually, Haggerty found a small company, Regency, which introduced the pocket radio just in time for the Christmas sales rush in 1954. More than 100,000 Regency portables were sold the first year, and pocket radio sales increased astronomically thereafter. The product was as successful for T.I. as for Regency, because it made the transistor a familiar object throughout the world. Some Texas Instruments people say Haggerty's multimillion-dollar crash program to put the transistor into a consumer product was actually aimed at a single consumer—Thomas Watson, Jr., the head of IBM. If that was Haggerty's aim, he hit a bull's-eye. Watson bought 100 Regency radios and distributed them among his engineers; according to an IBM executive, Watson told his people that "if that little outfit down in Texas can make these radios work . . . they can make transistors that will make our computers work, too." In 1957, Watson signed a purchase order that made Texas Instruments a key IBM supplier and provided a huge new market for T.I.'s transistors.

Eleven years after he had launched the radio, Haggerty got to thinking about the future of the integrated circuit. The chip was winning itself a niche in military and industrial markets but had yet to crack the computer industry or make even the smallest dent in the consumer market. Haggerty knew, and his engineers knew, that the chip represented a revolutionary advance in technology and that the product's reliability, capacity, and value were increasing every year—but the world at large did not know that. What was needed, Haggerty decided, was a dramatic demonstration of the benefits that monolithic circuitry could provide. And he had a particularly dramatic demonstration project in mind.

On an airplane trip with one of his engineers in the fall of 1965, Haggerty started talking about his belief that tiny, inexpensive, but complex circuits in integrated form would lead to new dimensions of pervasiveness. Why, the day would come, he said, when the chip would be built into a wide range of consumer products—when there would be integrated circuits in every home. As a matter of fact, he had been thinking about a consumer product that would be a perfect vehicle for the monolithic circuit. Before the plane landed, Haggerty had ordered the engineer to invent something wholly new: a miniature calculator that would fit in the palm of a hand—much lighter, much smaller, and much cheaper than any calculating machine that had ever been thought of before.

The man on the receiving end of that order was one of the most respected and fastest-rising engineers in the company—Jack Kilby. Since his invention of the integrated circuit during his first month at Texas Instruments, Kilby had received a series of raises and promotions befitting an employee who had given the company a firm position on the leading edge of monolithic technology. He had gone from simply "Engineer" to "Manager of Engineering" to "Manager, Integrated Circuits Development" to "Deputy Director, Semiconductor Research and Development Laboratory." Best of all, from Kilby's point of view, the company had essentially left him alone to do the kind of work he liked best—solving technical problems. When Haggerty came up with the formidable problem of building a pocket calculator, Jack Kilby was the obvious person to take it on.

"I sort of defined Haggerty's goal to mean something that would fit in a coat pocket and sell for less than $100," Kilby recalls. "It would have to add, subtract, divide, and multiply, and maybe do a square root, it would have to use some sort of battery as a power supply, to make it portable, and it couldn't be too much heavier than a fairly small paperback book." It was a tall order. Today, of course, the calculator Kilby described is utterly commonplace, but in 1965 it was something quite unprecedented. There were calculating machines available then, but none came close to meeting Haggerty's terms. The standard electronic desk calculator was as large and as heavy as a full-size office typewriter. It contained racks and racks of electronic parts and dozens of feet of wire. It ran on 120 volts of electricity and cost roughly $1,200—about half the price of a family car.

The president's own brainchild naturally became a matter of some priority at Texas Instruments. The company, which yields nothing to the CIA when it comes to secrecy, put Kilby in a shrouded office and told him always to refer to his new project by a code name. An earlier T.I. research program had been called Project M.I.T., so Kilby took the logical next step and named his effort Project Cal Tech. "It was a miserable choice," Kilby recalled afterward. "Anybody who heard it would have figured out that we had a crash project going on calculator technology." In any case, Cal Tech soon grew into a team effort. Kilby started looking around the semiconductor lab for engineers who would not be daunted by the sizable technical problems involved in inventing a new species of calculator. Among others, he settled on a friendly, easygoing young Texan named Jerry Merryman.

Merryman, who had come to Texas Instruments two years earlier at the age of thirty-one (and is still there today), represents a vanishing breed in high-tech industry—the self-taught engineer. After finishing high school in the country town of Hearne, Texas, he floated around and through Texas A&M for a few years but never stayed in place long enough to establish a major, much less earn a degree. Instead, he learned electrical engineering on odd jobs here and there and developed an almost intuitive sense for circuitry. "He's one of these guys who looks inside for a minute or two and then says, 'Well, maybe if we put that green wire over *there*,' " Kilby says. Like Kilby, Merryman has a fundamental confidence that any technical problem can be solved. "I just know," he says, "that you're going to find an answer if you think about it right. Eventually it'll come. A lot of inventions just happen on the way to work."

It was almost an act of faith in 1965 to believe that you could reduce the size, weight, and cost of an electronic calculator by factors of ten or more. One of the first things Kilby realized was that tearing apart existing adding machines to see how they worked—a process known as "reverse engineering"—would offer little, if any, help, because the basic architecture of this pocket-sized device would have to be completely new. And so the team started at ground zero, setting down the fundamental elements that their calculator would require.

In accordance with the architecture worked out by Alan Turing and John von Neumann, all digital devices, from the most powerful mainframe "Supercomputer" to the simplest handheld electronic game, can be divided into four basic parts serving four essential functions:

Input: The unit that receives information from a human operator, a sensory device, or another computer and delivers it to the processing unit. For the calculator, this meant a keyboard.

Memory: The unit that holds data—numbers, words, or instruction code—until the processing unit is ready to receive it.

Processor: The central control circuit that transfers data to and from various memory segments and manipulates the data. In a calculator, the processor performs the arithmetic.

Output: The display unit that shows the results of each calculation.

In addition to these four basic sections, Kilby also had to worry about something that is no problem on nonportable electronic devices—a power supply. A calculator meant to be carried in a pocket and used anywhere could not be designed for the 120 volts of power available from a wall socket. Instead, the calculator would have to operate from a battery; thus it would be limited to about 5 volts.

Because Haggerty was in a hurry, and because Kilby had a basic confidence that the job could be done, they decided on an all-points attack: the group would work on all its problems at once and hope that everything came together in the end. The input section—basically, the design of a small, power-efficient keyboard—was assigned to an engineer named James Van Tassel. Kilby himself took on the output section and the power supply. The memory and central processor—the calculator's electronic innards—were Merryman's responsibility.

"The basic rule was, everything had to be smaller than you'd ever thought you could make it," Merryman said later. "Now, one thing we did, we reduced the whole memory down to a single register. ["Register" is computerese for a short chain of transistors that can store a dozen or so binary digits.] The only problem with that was you'd need a lot of wires coming into that register, and there wasn't much room inside that case for a lot of wires." To solve that, Merryman designed a special "shift register"—a storage circuit that is laid out something like a large auditorium with only a single narrow aisle for people to come in and out. "That way, all the bits could come in there, sort of march in single file, and since they come in one at a time you get a lot of numbers in with just one wire."

Merryman's most serious problem, though, was with the logic circuitry. The desk-top electronic calculators of that time employed thousands of logic gates—AND, OR, and NOT circuits—to carry out binary calculations. "We were trying to build this whole circuit with only three bars," Merryman says ("bar" is the engineer's slang term for a silicon chip). "That left me about 400 gates in all—maybe 4,000 transistors. And I worked out a processing unit that only needed 400 gates. It almost worked, too—except it never could figure out how to get the decimal point in the right place. That took a whole bunch of extra gates." In the end, the team had to settle for a four-chip design with a total of 535 logic gates.

Meanwhile, Van Tassel had developed a working model for the keyboard, and Kilby had found a rechargeable battery that would run the device for three hours before running down. Memory, processor, input, and power supply were in good shape. That left only one problem—the output unit, for displaying the answer. But this proved to be an unusually thorny problem.

Contemporary desk calculators used a cathode ray tube—a miniature television set—for display, but such a system was far too heavy, fragile, and power-hungry for a portable machine. For a while, Kilby had hoped to use a row of tiny neon lights to display the answers; as it turned out, that system required at least 40 volts—out of the question. Just down the hall from the Cal Tech team, a T.I. researcher was working on a new electronic device—a "light-emitting diode," or LED—that was supposed to shine with a bright colored light when a minute current passed through it. This technology did, in fact, become the standard display technique for calculators and watches a few years later; in 1965, though, the diodes were not yet emitting much light.

There was nothing to do but invent something new. So Kilby developed a thermal printing technique, in which a low-power printing head would "burn" images into heat-sensitive paper; the idea worked perfectly, and the process is still widely used in low-cost, power-efficient printers.

All this activity consumed a little more than twelve months. One day late in 1966, Merryman recalls, "the thing was all laid out on the table like a person spread out on an operating table, all split open, wire running all over, and we punched in a problem, and it worked!" Silently, and almost instantaneously, the right answer came spinning out of the machine on a strip of paper tape. The Cal Tech group took their prototype in to Haggerty, who nodded with satisfaction—and called in the patent lawyers. It took another year before the design was perfected and the patent application—for a "Miniature Electronic Calculator"—could be filed. Although handheld calculators have come a long, long way since then, the Cal Tech team's architecture is still the gist of all such devices; even today most T.I. models carry the number (3,819,921) of the original Kilby-Merryman-Van Tassel patent.

The electronics of the new device were so far ahead of their time that it took years to turn out the initial production models. The world's first pocket calculator, the Pocketronic, was not introduced until April of 1971—April 14, to be exact (the marketing people thought they might win the attention of taxpayers working late on Form 1040). By today's standards, that first model was a dinosaur—a four-function (add, subtract, multiply, divide) calculator that weighed 2½ pounds and cost about $150. But it sold like crazy.

You would need a fairly high-powered calculator to keep track of what happened next. Five million pocket calculators were sold in the United States in 1972. As new features were added and prices plummeted—a four-function model cost $100 by the end of 1972, $25 in 1976, and $10 in 1980—sales doubled year after year. To borrow a word from Patrick Haggerty, the pocket calculator became pervasive. Within a decade after the first pocket calculator was sold in the United States, the country had more calculators than people. As Haggerty had predicted, the new microelectronic gadget created a market that had simply not existed before. Tens of millions of people who never considered purchasing an adding machine or a slide rule decided they wanted to own a pocket calculator. "How many housewives actually need to know the square root of a number?" wrote Ernest Braun and

Stuart MacDonald, two English scholars who analyzed the phenomenon. "But then, the technology is ridiculously cheap. For a fraction of the cost of one week's housekeeping, one can have permanent access to any number's square root."

Today, with a four-function calculator—a model the industry calls "plain vanilla"—available for $6.95 or so, the U.S. market is virtually saturated. Yet Americans still buy between 26 million and 30 million replacement calculators each year. Worldwide, the calculator market is a billion-dollar-per-year business, with sales approaching 100 million calculators each year.

About ten years after its birth—unusually soon for a major consumer product—the calculator reached what retailing experts call the "mature end" of the product cycle. Among other things, that means the market is broken into clearly segmented categories—four of them, in the calculator's case. At the high-price end, where prices still exceed $125, are the programmable calculators, which fit somewhere along the fuzzy line between calculators and computers. In between, at prices ranging from $30 to $130, are the scientific, or "slide rule," calculators (which can handle the sophisticated math involved in science and engineering) and business calculators (which turn out amortization schedules, net present values, and the like). The most interesting facet of today's calculator market is the bottom, where makers compete in sheer ingenuity to spice up the plain vanilla calculator with exotic toppings.

If you set out tomorrow morning to buy a simple four-function calculator, you would have your choice of denim-clad calculators, designer-label calculators, wristwatch-sized calculators, ring-sized calculators, or calculators built into pens and pendants. You could pay anywhere from $6.95 for basic black at K Mart to $7,500 for a 24-karat gold version at Pierre Cardin. Casio, the Japanese calculator firm, provides a calculator built into an electronic musical keyboard ("the first piano in history that can compute a square root," Casio boasts). Texas Instruments, still the leading American manufacturer, is fighting back with calculators that talk to you—"with a warm, synthesized voice," the company's catalogue promises. There are calculators on the market today that read your star chart, count your calories, track your biorhythms, translate Spanish (or French, or German), convert dollars to yen (or pounds, or rubles), or let you compose a song or conquer an electronic Rubik's Cube if you get bored between calculations. There are calculators designed specifically for securities analysts, accountants, pilots, navigators, handicappers, and—naturally—computer programmers.

Another consumer application of the chip, born the same year as the handheld calculator and just as "pervasive" now, was the electronic, or digital, wristwatch. As a technical matter, the digital watch is markedly easier to make than a calculating machine. It is based on a convenient natural phenomenon called "crystal oscillation," which is the physicist's way of saying that pure crystals of certain elements will oscillate, or vibrate back and forth, when connected to a source of electric current (e.g., a small battery). The rate of vibration depends on the atomic structure of the element; for a given material, though, the rate never varies. Certain crystals of the common mineral quartz, for example, will vibrate back and forth, back and forth, precisely 3,579,545 times each second.

A precise oscillator, be it the 5-foot brass pendulum of a grandfather clock or the .5-cm. flake of quartz in a wristwatch, is the heart of any timepiece. All the watchmaker needs is a mechanism to count the back and forth oscillations—and counting is one of the simple tasks that binary logic gates can perform. In the digital watch a logic gate called a "JK flip-flop" counts the vibrations of the crystal. Every time the count hits 3,579,545, the gate sends a pulse to the display unit and the watch records the passage of another second. Another set of gates on the same chip counts 60 seconds and updates the minute display; another counts minutes to update the hour. If you tore apart your digital watch (why not? you can get a new one for $5.95), you would find, in place of the gears, springs, bearings, and bushings of a traditional timepiece, only four parts: a battery, a crystal, a chip, and the display unit.

Nonetheless, the first digital watch—produced by an American firm under the "Pulsar" brand name and introduced in the fall of 1971—was marketed as a decidedly high-bracket item. The 24-karat gold Pulsar was priced at $2,000; a stainless steel model cost $275. Characteristically, as the electronic watch improved—getting smaller, easier to read, more power-efficient—its price fell. By its tenth birthday, the digital watch, festooned with all manner of whistles, bells, and baubles, was down to an average price of $40 and was selling at the rate of 50 million units per year.

The American reader may have spotted in this history a perfectly legitimate excuse for chauvinism. Despite the predominance of names like Sony and Seiko, Canon and Casio, the major consumer products of the microelectronic age all resulted from pure Yankee ingenuity, as did the fundamental breakthrough—The Monolithic Idea—that made such advances possible in the first place. Many, perhaps most Americans today consider the microelectronic revolution just another import from Japan—one more manifestation of the Japanese genius for technology and marketing. In fact, the flow of genius has gone in the other direction. The history of microelectronics has been a history of Japanese firms—and other companies around the world—learning at the feet of American innovators. This familiar pattern was played out once again with the development of the device that has taken microelectronics further than ever down the path of pervasiveness—the microprocessor.

The story of the microprocessor begins in Tokyo, but the scene shifts rapidly to Silicon Valley. In 1969 a Japanese business-machine manufacturer, Busicom, was planning a new family of desk-top printer-calculators but could find no engineers in Japan capable of designing the complex set of integrated circuits the machines would require. Busicom sought help—from Bob Noyce, who was still putting together his new company, Intel. The Japanese signed a contract with Intel calling for design and production of twelve interlinked chips for the new line of machines. Busicom sent a team of engineers to Intel to oversee the work. Noyce, meanwhile, handed the problem to a one-man team—Marcian E. "Ted" Hoff, a thirty-four-year-old Ph.D. who had been lured away from a teaching job at Stanford by the prospect of broader horizons in industry. Although Hoff was an expert on micro-circuits, his real ambitions were somewhat larger. He had always wanted to design his own computer.

When the Busicom engineers showed Hoff their tentative plans for the twelve chips they needed, the American was appalled. The arrangement was outrageously complex—some of the simplest functions would require sending the same number into and out of two or three different memory registers—and could not possibly be implemented at an acceptable price. Moreover, it just seemed wasteful to put dozens of man-years into design of a set of circuits that could be used only in one small group of machines.

This last concern was important to Noyce as well. By the end of the sixties Noyce was worried about the rapid proliferation of different integrated circuits, each designed for its own special purpose. Every customer who wanted a chip for his product was demanding a custom-designed chip just for that product. "If this continued," Noyce and Hoff wrote later, "the number of circuits needed would proliferate beyond the number of circuit designers. At the same time, the relative usage of each circuit would fall. . . . Increased design cost and diminished usage would prevent manufacturers from amortizing costs over a large user population and would cut off the advantages of the learning curve."

Looking ahead, Noyce saw that the solution to proliferation of special-purpose integrated circuits would be the development of general purpose chips that could be manufactured in huge quantities and adapted ("programmed") for specific applications. Hoff was intrigued by this insight and decided to apply it to his Busicom assignment. He came up with a strikingly new design for the Japanese: a general-purpose processor circuit that could be programmed for a variety of jobs, including the performance of arithmetic in Busicom's machines. As Hoff pointed out, this approach would permit much simpler circuitry than the Japanese firm had suggested. Indeed, by the summer of 1971, Hoff was able to put all the logic circuitry of a calculator's central processor unit, or "CPU," on a single chip. The CPU could be coupled with one chip for memory, one for storage registers, and one to hold the program; the entire family of calculators would require only four integrated circuits. In his job as a circuit designer, Hoff had, in fact, fulfilled his personal dream: he had designed his own general-purpose computer.

By the summer of 1971, though, the calculator industry was in the throes of great change. The introduction of Jack Kilby's $150 hand-held calculator that spring had changed the rules of the game; companies like Busicom with their heavyweight $1,000 machines were in big trouble. Accordingly, Busicom told Intel it could no longer pay the price originally agreed upon for the new chips. Negotiations ensued. Busicom got its lower price but in return gave up its exclusive right to the chips. Intel was now free to sell Hoff's general purpose CPU-on-a-chip to anybody.

But would anybody buy it? That question spurred furious debate at Intel. The marketing people could see no value in a one-chip CPU. At best, a few minicomputer firms might buy a few thousand of the chips each year; that wouldn't even pay for the advertising. Some directors were worried that the new circuit was too far afield from Intel's real business. Intel, after all, was a *circuit* maker; Hoff's new chip was a single circuit, all right, but it really amounted to a complete *system*—almost a whole computer. There was strong pressure, Noyce and Hoff wrote later, to drop the whole thing.

Intel had recently hired a new marketing manager, Ed Gelbach, and he arrived at the company in the midst of this controversy. As it happened, Gelbach had started in the semiconductor business at Texas Instruments; like everyone else at T.I., he was steeped in Patrick Haggerty's view of the world. Gelbach realized immediately that Intel had reversed the course of the industry by producing a general-purpose chip. "General purpose," Gelbach saw, was just another way of saying "pervasive." The real markets for the new device, he said, would be completely new markets; with this one-chip central processor—known today as a microprocessor—the integrated circuit could "insert intelligence into many products for the first time."

And so Intel's new "4004" integrated circuit went on sale for $200 late in 1971. With mild hyperbole, Intel advertised the device as a "computer on a chip." Gradually, as people realized that it really could work just about anywhere, the microprocessor started showing up just about everywhere. A typical application was the world's first "smart" traffic light. It could tell, through sound and light sensors, when rush hour was starting, peaking, or running down; the tiny CPU would alter the timing of red and green in response to conditions to maximize traffic flow. Soon there was a smart elevator, a smart butcher's scale, a smart gas pump, a smart cattle feeder, a smart typewriter, and a bewildering array of other "smart" devices.

Texas Instruments, of course, was hardly pleased to let its arch-rival steal a march on such an important new battleground. Working on a contract for a customer who wanted circuitry for a "smart" data terminal—a keyboard-screen combination that could communicate with large computers far away—a T.I. engineer named Gary Boone developed a slightly different version of a single-chip processor unit. Boone's version, called the "TMS 1000," received the first patent awarded for a microprocessor. Then, at the end of 1971, Boone and another Texas Instruments engineer, Michael Cochran, produced the first prototype of an integrated circuit that actually was a computer on a chip. The single monolithic circuit contained all four basic parts of a computer: input circuits, memory, a central processor that could manipulate data, and output circuits.

A year later, Intel came out with a second-generation microprocessor; since it had roughly twice the capacity of the original 4004 chip, the new device was called the "8008." The introductory price was $200. Today, of course, that device has been considerably improved; the current version, known as an "8080," sells for about $3.00 in volume purchases. This new edition was even more successful than the 4004, and its success spawned a raft of imitators. In the late 1970s dozens of new companies—outfits with names like Signetics, Synertek, MOS Technology, Advanced Micro Devices, and Zilog—sprang up to produce their own versions of the microprocessor.

It was the marriage of the microprocessor and a group of devices called "transducers" that finally brought microelectronics into every home, school, and business. A transducer is an energy translator; it converts one form of energy into another. A telephone receiver is a transducer, changing your voice into electrical pulses that travel through the wire. The keyboard on a calculator converts physical pressure from a finger into pulses that the central processor can understand. Other sensors can turn sound, heat, light, moisture, and chemical stimuli into electronic impulses. This information can be sent to a microprocessor that decides, according to preprogrammed directions, how to react to changes in its environment.

A heat-sensitive transducer can tell whether a car's engine is burning fuel at peak efficiency; if it is not, the transducer sends a pulse to logic gates in a microprocessor that adjust the carburetor to get the optimum mixture of fuel and air. A light-sensitive transducer—the familiar "electric eye"—at the checkout stand reads the Universal Product Code on a carton of milk and sends a stream of binary pulses to a microprocessor inside the cash register. The central processor queries memory to find out the price assigned to that specific product code today, adds that price to the total bill, and waits patiently (this has all taken three-thousandths of a second) for the transducer to read the next product code. A moisture sensor and a heat sensor inside the clothes dryer constantly measure the wet clothes and adjust the machinery so that the laundry will be finished in the shortest possible time.

As if this weren't enough, microelectronics will soon be at work inside the human body. A microprocessor that controls a speech-synthesizing chip can be connected to a palm-size keyboard that permits the mute to speak. Now under development is a chip that may be able to turn sound into impulses the brain can understand—not just a high-tech hearing aid but an electronic ear that can replace a faulty organic version. Other experiments suggest the possibility of an implantable seeing-eye chip for the blind—a light-sensitive transducer connected to a microprocessor that sends intelligible impulses to the brain.

Among the countless new applications that people dreamed up for the "computer on a chip" was, of all things, a computer—a completely new computer designed, not for big corporations or public agencies, but rather for ordinary people. The personal computer got its start in the January 1975 issue of *Popular Electronics* magazine, a journal widely read among ham radio buffs and electronics hobbyists. The cover of that issue trumpeted a "Project Breakthrough! World's First Minicomputer Kit to Rival Commercial Models." Inside, the reader found plans for a homemade "microcomputer" in which the Intel 8080 microprocessor replaced hundreds of individual logic chips found in the standard minicomputer of the day. The Popular Electronics kit was strictly bare-bones, but it gave anybody who was handy with a soldering iron the chance to have a computer of his own for a total investment of about $800. At a time when the smallest available commercial model sold for some $30,000, that was indeed a breakthrough. Readers sent in by the thousands. The other electronics magazines started offering computer kits of their own. Within a year thousands of Americans were tinkering with their own microprocessor-based personal computers.

These early computer buffs—"addicts" might be a more descriptive term—began forming clubs where they could get together for endless debates about the best approach to bit-mapped graphics or the proper interface for a floppy disk or the relative merits of the 8080, 6502, and Z80 microprocessors. At one such organization in Silicon Valley—a group called the "Homebrew Computer Club"—two young computerphiles, Steven Jobs and Stephen Wozniak, convinced themselves that there had to be a larger market for personal computers than the relatively small world of electronics tinkerers. The gimmick, they decided, was to design a machine that was pretty to look at and simple to use. The important thing was that the personal computer could not be intimidating; even the name of the machine would have to sound congenial. Eventually, Jobs settled on the friendliest word he could think of—"Apple." It had nothing to do with computers or electronics—but that was the whole point. The two started a computer company called Apple. By its fifth birthday Apple had sales of $139 million and a host of competitors.

There was a time—when computers were huge, impossibly expensive, and daunting even to experts—when the sociological savants regularly warned that ordinary people could become pawns at the hands of the few corporate and governmental Big Brothers that could afford and understand computers. This centralization of power in the hands of the computer's controllers was a basic precept of Orwell's 1984. The real-life version of 1984, in contrast, was marked by the mass distribution of microelectronics—and a resulting decentralization of computing power. In the real 1984, millions of ordinary people could match the governmental or corporate computer bit for bit. In the real 1984, the stereotypical computer user had become a Little Brother seated at the keyboard to write his seventh-grade science report.

Three decades after he began proselytizing for the concept of "pervasiveness," Patrick Haggerty had been proven right. Microelectronics did pervade nearly every aspect of society, replacing traditional means of control in familiar devices and creating new aspects of human activity that were previously unknown. By shrinking from the room-sized ENIAC to the pinhead-sized microprocessor, the computer had imploded into the basic fabric of daily life.

Haggerty lived until 1980, long enough to see his prediction coming true but not to determine what the impact would be. His successors are just now trying to grapple with that issue. Prominent among those who are fascinated with the effect of microelectronics on human society is one of the patriarchs, Robert Noyce. "Clearly, a world with hundreds of millions of computers is going to be a different world," he said not long ago. "But what will come of it? Who can use all that intelligence? What will you use it for? That's a question technology can't answer."

DIM-I

A SMALL BUT NOTEWORTHY segment of American industry, dead of competitive causes, was formally laid to rest in Washington, D.C., on a summer day in 1976. The rite of interment was a small ceremony at the Smithsonian Institution in which Keuffel & Esser Company, the venerable manufacturer of precision engineering instruments, presented the museum with the last Keuffel & Esser slide rule—together with the milling machine the firm had used to turn out millions of slide rules over the years for students, scientists, architects, and engineers. Shortly past its three hundredth birthday—it had been invented by the seventeenth-century British scholar William Oughtred, whose other great contribution to mathematics was the first use of the symbol × for multiplication—the slide rule had become a martyr to microelectronic progress.

"Progress," in this case, meant Jack Kilby's handheld calculator. In the five years after the small electronic calculator first hit the market, K & E, the largest and most famous slide rule maker on earth, had watched sales fall from about 20,000 rules per month to barely 1,000 each year. By the mid-seventies, even Keuffel & Esser was selling its own brand of electronic calculator; the slide rule had become a museum piece. "Calculator usage is now 100 percent here," an M.I.T. professor told *The New York Times* in 1976.

The last slide rule was buried in a bin on a storage shelf at the

Smithsonian's National Museum of American History. Someday it may be dusted off and put on display; at present, a special appointment must be made to view the relic. The curators say nobody ever bothers. Still, the slide rule lives on, in the affectionate memory (and frequently in the desk drawers) of a whole generation of scientists and engineers—including Jack Kilby. Kilby and his contemporaries recall the slide rule today with the same fond regard that an old golfer holds for a hickory-shafted mashie niblick or an old auto buff reserves for the 1936 Cadillac V-12. In a requiem for the slide rule published in the journal *Technology Review,* Prof. Henry Petroski recalled that the Keuffel & Esser Log-Log Duplex Decitrig he bought as an undergraduate became his most valued possession: "That silent computational partner [was] my constant companion throughout college and my early engineering career."

From a practical viewpoint, the competition between the slide rule and the calculator was completely one-sided from the beginning. The slide rule was really just a calculator's assistant; it could provide quick, approximate answers to multiplication and division problems, but the trickier parts of the problem—determining the order of magnitude and putting the decimal point in the right place—were left to the human operator. The most primitive pocket calculator, in contrast, solved problems precisely, down to the last decimal, with speed and accuracy that no slide rule, not even K & E's finest, could match. Today, though, in the hazy afterglow of nostalgia, engineers tend to look back on the slide rule's drawbacks and see virtues. "The absence of a decimal point," Petroski wrote, "meant that the engineer always had to make a quick mental calculation independent of the calculating instrument to establish whether the job required 2.35, 23.5, or 235 yards of concrete. In this way, engineers learned early an intuitive appreciation of magnitudes. Now the decimal point floats across the display of an electronic [calculator] among extended digits that are too often copied down without thought."

The slide rule was a simple instrument, nothing more than a wooden ruler that slid back and forth in a calibrated wooden frame. To hear the engineers tell it now, that simplicity was part of its charm. "It has sort of an honesty about it," Jack Kilby says. "With the slide rule, there're no hidden parts. There's no black box; there's nothing going on that isn't right there on the table." To put it another way, the slide rule was not threatening. Nobody ever called the slide rule a "mechanical brain." Nobody ever said the slide rule was imbued with something called "artificial intelligence." There were no movies about runaway slide rules plotting to dominate mankind. The slide rule, hanging at the ready in its holster from the belts of Fermi, Wigner, and Von Braun, helped men create the first nuclear chain reaction and send rockets to the stratosphere. But the rule was always recognized as nothing more than a tool. It was no more "intelligent" than a yardstick or a screwdriver or any other familiar tool. Like the yardstick and the screwdriver, the slide rule was just an ignorant mechanism.

Someday—fairly soon, probably, given the accelerated pace of technological development—the pocket calculator, the desk-top computer, and other digital marvels of our day will themselves be museum pieces, on exhibit in a gallery called "Primitive Microelectronic Tools" or some such. As we file past with our grandchildren, we may well break into nostalgic smiles of fond regard for these devices that used to be considered so revolutionary. The kids, no doubt, will be amazed to learn that, back in the seventies and eighties, people resisted those simple tools, resented them, even feared them—feared that digital computers, known by the melodramatic, but erroneous, label "electronic brains," might replace poor bungling man as the ruling intelligence on Earth. To our grandchildren's generation, how foolish this will seem! Why, they will wonder, would anyone have feared an ignorant mechanism?

To most of us today, of course, even the simplest digital calculator seems formidable. It is, as Jack Kilby suggested, a black box. What goes on inside the black box is, for most people, black magic. You push the keys. The answer shows up on the screen. The rest is mystery. It's no wonder that calculators, computers, and their ilk are thought of as "intelligent" machines; what other explanation could there be? In fact, the explanation of how a calculator gets the answer is not at all magical. The "magic" inside the black box really involves a series of mathematical and logical techniques carried out by artful arrangements of electronic switches arrayed in logic gates. Electronic impulses are pulled this way and that through the maze of electronic switches by blind physical force.

The electrons racing through a logic chip have as much "intelligence" as water running down a hill. Gravity pulls the water. Electricity

—the attraction and repulsion of electric charges—pulls the electrons. If men build sluice gates and irrigation canals in the right combinations, they can make surface water flow where it's needed. If men build logic gates and connecting leads in the right patterns, they can make electronic pulses flow where they're needed to solve the problem. In each case, the mechanism does the work, but in each case, it's an ignorant mechanism; the human designer provides all the intelligence.

If you punch into your calculator the task of adding 3+2, the mechanism will produce the answer 5. The calculator gets the answer not because of "artificial intelligence" but rather because a genuine intelligence—the human mind—has designed the mechanism so that it has to produce the right answer.

Using switching logic (from the minds of George Boole and Claude Shannon) implemented by transistors (from the minds of Shockley, Bardeen, and Brattain) contained in a monolithic circuit (from the minds of Kilby and Noyce), the humans who build digital machines have designed an addition circuit in such a way that the pattern of pulses representing binary 5 is the only possible combination that can come out when binary 3 and binary 2 are put in. To get that sum, however, out of a mechanism consisting entirely of switches turning on and off, off and on, the humans have had to go to extreme lengths. For a machine as dim-witted as a computer to solve 3+2, the problem must be broken down into an absurdly detailed sequence of instructions that lead the machinery through its paces, step by elementary step. If there is magic in a pocket calculator, it is not in the machinery; it is in the humans who had the wit and the patience to program the machine to do its job.

To bring the point home, we can take a guided tour through the interior of the black box and watch a typical digital mechanism from the inside as it does its stuff. We'll look at a simple pocket calculator—so simple it exists only in the pages of this book—called the "DIM [Digital Ignorant Mechanism]-I." The design of DIM-I to be set forth here is based on the familiar four-function calculator available anywhere for no more than $15. To a considerable extent, though, the basic architecture of a $15 calculator is the same as that in a $150 video game, a $1,500 personal computer, and the $15 million supercomputers that guide NASA's rockets through the cosmos. The bigger, more expensive machines can handle more information and perform a larger variety of functions, but the *modus operandi* is the same. If you've seen one digital machine at work, you've pretty much seen them all.

On the outside, DIM-I is in fact a black plastic box. The box has eighteen keys, one each for the digits 0 through 9 and eight others for various functions such as "+," "=," and the like. It also has a display screen that can show numbers up to eight digits long. Pry open the black box, and you'll find mostly nothing; a largely empty space containing a tiny battery, a few wires—perhaps a dozen or so—and a printed circuit board on which there sits another, smaller black box. This one is a black piece of plastic that looks like a man-made millipede: an inch-long rectangle with a symmetrical array of copper legs sticking out from each side. That millipede is the chip—or, more precisely, the plastic package that holds the chip. The two rows of legs are the leads that connect the keyboard and the display screen to the chip.

Inside a computer—even the smallest personal computer—there are whole platoons of these small black millipedes, each chip designed for a specific function. Open the back of an IBM personal computer, for example, and you will see about 200 separate chips. There are memory chips and logic chips and micro-processor chips, all lined up in formation on the circuit boards. A simple calculator, in contrast, employs a single chip that has various different functional circuits built into it.

The chip inside DIM-I is a TMS 1000C, one of the most common microprocessor chips for simple calculator applications. It has a set of logic gates that reads electronic signals from the keyboard and "encodes" them into the binary form the calculator can understand. It has another set of gates that "decodes" binary information back into decimal form and sends it to the display screen. It has a relatively small number of memory units—chains of transistors lined up in ordered rows so that each one can be separately addressed, allowing "random access." It has an arithmetic processor unit—a group of transistors arranged in gates so they can use Boolean logic to perform simple math. In other words, the chip has the necessary circuitry to do four things:

1. Sense numbers punched into its keyboard
2. Write them on an electronic scratch pad so it doesn't forget
3. Add, subtract, multiply, or divide them

4. Report the sum in the form of lighted digits on its display screen

Which is just another way of saying that DIM-I can carry out the four essential jobs of any digital device:

1. input
2. memory
3. processing
4. output

A fancier machine—say, DIM-II or DM-III—might be able to handle larger numbers, find percentages or square roots, and store greater quantities of information in memory. By the time one gets to DIM-X or DIM-XXV, the machine can deal with words as well as numbers and manipulate the information in countless ways that are beyond the wildest dreams of a little calculator. DIM-I by comparison, can't do much, but it can do just enough.

The processing circuitry of DIM-I, like that in any digital device, also has a central set of logic gates called the "Control Unit"; this is a sort of central switchboard that busily directs electronic pulses here and there, from input to memory to processor to the display screen, as needed to solve the problem. The Control Unit is itself controlled, in turn, by a simple, familiar mechanism that is essential to the operation of any digital machine—a clock.

To the extent that we attribute anthropomorphic characteristics at all to computers, the proper analogy would be, not that the machine has a brain, but that it has a heart—a steady, pulsing central rhythm instrument that orchestrates and controls everything that happens. The computer engineers call the central clock a "clock generator" because it is really a circuit that generates pulses at a perfectly steady, unvarying rate. The clock is to the computer what the bandleader is to the band; it stands there keeping time, ONE-two-three ONE-two-three, so that each part will come in at the right beat.

The clock inside DIM-I beats, without variation, every 10 millionths of a second—that is, it emits 100,000 pulses every second, or one pulse every 1/100,000 second. This is a fairly standard clock rate for cheap, simple pocket calculators. It is slow compared to the operating speed of large mainframe computers—or small personal computers, for that matter—but extremely fast compared to, say, a flash of lightning or the blink of an eye. An eye's blink takes about 0.30 second; in that time the clock inside a simple calculator will have pulsed 30,000 times.

There is no way for humans, in our pokey world of seconds, minutes, hours, to conceive of a time period like 1/100,000 second, much less the microsecond (1/1,000,000 second), the nanosecond (1/1,000,000,000 second), the picosecond (1/1,000,000,000,000 second), or the femtosecond (1/1,000,000,000,000,000 second). On the human scale, anything that lasts less than about a tenth of a second passes by too quickly for the brain to form a visual image and is thus invisible; if the duration is less than a thousandth of a second or so, the event becomes too fast even for subliminal perception and is completely outside the human sphere. The speed of microelectronic events puts them in a world far removed from the human realm; how can any human contemplate a thousandth of a thousandth of a second? Computer engineers, practical types not given to metaphysical speculation, don't even try. They just become, as Bob Noyce says, "reconciled" to the thought that their machines work at unthinkably high speeds.

The inconceivable speed of operation comes about because the "moving parts" of a digital machine are electronic pulses that travel inconceivably fast over distances inconceivably small. An electric charge moves at the speed of light, 186,000 miles per second. This makes for extraordinarily rapid transit, even at transcontinental distances.

A baseball fan in Boston turns on his TV to watch the big game in Los Angeles. He watches the pitch coming in, the hitter starting to swing, and the smack of bat against ball. He does not see all this at the precise instant it happens, however; the electronic signals carrying the picture to his television take about 0.016 second to travel the 3,000 miles from coast to coast. Thus the viewer will not see the bat hit the ball until about 1/62 second after the impact actually occurs.

The signals that turn switches on and off inside a calculator move at the same speed, but they do not have to travel 3,000 miles. On a quarter-inch-square integrated circuit containing 10,000 components, individual transistors are spaced a few ten-thousands of an inch apart.

Electronic charges moving at the speed of light traverse those distances in very short periods of time. Consequently, the clock generator that controls the dispatch of pulses around the chip can be set to tick at very short intervals.

In setting the clock rate for digital machines, however, the human designers have to consider not only the transit time for pulses racing from one transistor to another but also the time it takes for each transistor to switch from on to off. The computer is made of a long, long chain of transistors; each one has to wait for the transistor ahead of it in line to flip one way or the other before it can do anything. In most modern computers the transistors' switching time (known to the engineers as "propagation delay") is a more serious constraint than the travel time for signals moving through the circuit.

"Propagation delay," like many other elements of microelectronic jargon, is a complex term for a simple and familiar phenomenon. Propagation delays occur on the freeway every day at rush hour. If 500 cars are proceeding bumper to bumper and the first car stops, all the others stop as well. When the first driver puts his foot back on the accelerator, the driver in the second car sees the brake lights ahead of him go off and he, in turn, switches his foot from brake to accelerator. The switching action is then relayed down the chain of cars. Even if each driver has a switching time from brake to accelerator of just one second, the last car will have to wait 500 seconds—about 8½ minutes—because of the propagation delay down the line of traffic.

The transistors inside a digital device go from brake to accelerator—from on to off—somewhat faster. In the most high-powered machines, the switching time is about one nanosecond—one billionth of a second. In the current generation of personal computers the transistors switch in about 0.5 microsecond—half a millionth of a second. By those standards a pocket calculator like DIM-I is a tortoise. Its transistors take about 5 microseconds to go from on to off. To account for that propagation delay and to allow a little additional time for signals to travel through the circuitry on the chip, the clock rate—the time signature that sets the tempo for each operation—in a small calculator is set at 10 microseconds, or one pulse every hundred-thousandth of a second.

The ticking of the clock inside a calculator or computer regulates a repetitive cycle of operations—the "instruction cycle"—that the machine performs continuously, over and over, as long as it is turned on. The instruction cycle is a two-stage affair.

On the first clock pulse the computer's Control Unit—the switchboard—sends a message to memory, asking for the next instruction. The instruction (in the form of a binary pattern of charges) flashes back to the switchboard—which holds it until the next tick of the clock. When the next clock pulse comes, the switchboard sends the instruction on to the appropriate part of the computer for execution. When one instruction has been executed, the controller waits for the next pulse. The clock ticks, and the controller fetches the next instruction; the clock ticks again, and that instruction is carried out.

This two-stage instruction cycle—fetch and execute, fetch and execute—is the vital rhythm of the computer's life, as fundamental and as constant as the two-stage respiratory cycle—inhale and exhale, inhale and exhale—of the human body. The clock generator regulates the process so that each signal—from memory, from the control unit, from the keyboard, from anywhere—arrives at its destination before the next signal starts its journey through the circuit.

In DIM-I, as in video games and other simple digital tools, the sequence of instructions—the "program"—is permanently installed in memory when the machine is built. Such preprogrammed devices can do the exact task they were built for, and nothing else. A computer, in contrast, is not restricted to built-in programs; its versatility comes from the fact that it can be programmed by each user to perform a broad range of tasks. This is the difference between a "universal" machine—a computer—and a "dedicated" tool like a calculator. Buying a calculator is like buying a ticket on the railroad: you can only go where the company's tracks will take you. If you have a computer, on the other hand, you can drive it wherever you want to go.

In any case, the program, whether built in at the factory or provided separately, is a long chain of instructions that reside in a designated block of memory. In order to do any job, no matter how mundane, the machine's central control unit has to fetch individual instructions from memory and execute them, one at a time, one after another.

The fundamental fetch-and-execute cycle begins the instant DIM-I is turned on. The clock generator starts generating regular pulses, beeping out a pulse every 1/100,000 of a second. If you turn on a calculator and then put it down to reach for a pencil or answer the phone, the calculator may look as though it is doing nothing. In fact, it is working furiously away, fetching and executing instructions —a special set of instructions that continually checks the keyboard, looking for "input." The special program any digital device follows when it is waiting for a human to press its keys is called the "idle routine." The idle routine is what every computer or calculator is engaged in most of the time.

DIM-I starts into the idle routine with the first clock pulse. *Tick*. The control unit goes to the first location in the memory bank and fetches the first instruction: "Check the '1' key on the keyboard." *Tick*. Now the controller executes that instruction; it sends an electronic query to the keyboard: "Has anybody pushed the '1' key lately?" The electronic reply comes back: "No." With this sequence—the first instruction was fetched and then executed—the first instruction cycle has been completed.

Tick. The idle routine continues. The control unit goes back to memory and fetches the second instruction: "Check the '2' key on the keyboard." *Tick*. The controller calls the keyboard: "Has anybody pushed the '2' key lately?" The electronic reply: "No." *Tick*. The controller fetches the next instruction: "Check the '3' key." *Tick*. The controller queries the keyboard: "Anybody pushed the '3' key lately?" Reply: "No."

At this point, even the most primitive intelligence would start to appreciate that the drill here is to check all the keys seriatim to see if any has been pushed. DIM-I, however, has no intelligence; it is an ignorant mechanism. It knows nothing but its unvarying fetch-and-execute routine.

Tick. Doggedly, the controller fetches the instruction in the next memory location: "Check the '4' key on the keyboard." *Tick*. Execute. *Tick*. Fetch. *Tick*. Execute. This is a computer's life. One by one, the control unit will check each of the eighteen keys. If none has been pushed, it runs through the whole routine again, and again, and again. Any intelligent creature would be driven to distraction after the first few repetitive milliseconds. A digital device, in contrast, races its way through billions and billions of fetch/execute cycles every working day, never getting bored, never getting tired.

At some point, every million instruction cycles or so, DIM-I's unceasing vigilance is rewarded: somebody pushes a key. The keyboard reports to the switchboard that a key has been pushed. But which key?

Each key on the keyboard of a calculator or computer is a switch, just like the light switch on a wall. When somebody flicks the light switch on the wall, it permits current to flow to the light bulb. When somebody pushes a key on the keyboard, it permits current to flow to the Control Unit. It is just a quick pulse of current—a swarm of electrons—and it is the same pulse no matter which key is hit. The "1" key, the "9" key, the "+" key—every key sends the same pulse. How, then, does the central Control Unit know which key has been pushed? That's where the clock comes in.

The Control Unit at the heart of DIM-I is like a stationmaster at some isolated depot along the main freight line. The stationmaster knows that four trains come through from the south every day: the 10 A.M. from Tulsa, the noon train from Natchez, the 2 P.M. from Texarkana, and the 4 P.M. from Fort Worth. His job is to watch for the trains and switch each one onto the right track to reach its destination.

Each morning at about nine forty-five the stationmaster hears the remote rumble of a distant engine chugging in. He can't see the train that is coming, and all trains sound alike at a distance. Yet the station man knows that this is the train from Tulsa. It has to be; it's the only train scheduled for that hour. He checks off the Tulsa train in his logbook; then he reaches in the desk drawer and fetches his stationmaster's manual to find the instruction telling which track he should route it to. At eleven forty-five he hears another engine; the stationmaster can't see that, either, and all trains sound alike. But he takes out the logbook again and checks off the train from Natchez. He knows it has to be the Natchez train, because the timetable tells him so.

So it is for the Control Unit. All through the idle routine it has been querying each of the eighteen keys, one at a time. Each key is queried at a specific time interval—intervals dictated by the steady ticking of the clock generator. The Control Unit can receive a pulse from a given key only when it is addressing that particular key. If a pulse comes in from the keyboard during the interval when the Control Unit is addressing the "3" key, then it had to be the "3" button that was pushed. Pulses from the keyboard, like locomotives at a distance, all sound alike. But the Control Unit knows this pulse had to come from the "3" key, because the timetable tells it so.

Except that there's one other possibility: the surge of current may not have come from the keyboard at all. It could have been nothing more than random electronic "noise." Our atmosphere is almost always noisy with stray electromagnetic pulses, some generated by nature and others by the machinery and weaponry of mankind. Weapons, and particularly nuclear weapons, are prolific producers of electromagnetic pulses—or "EMP," to use the Pentagon acronym. One of the many unknowns about modern nuclear arsenals is the quantity of EMP that might result from an atomic blast. It is considered possible that any nuclear weapon, once triggered, would release so much electromagnetic "noise" that the control units of the world's computers—including the computers supposedly directing both sides' strategic planning—would be flooded with pulses and would break down in a state of terminal confusion. Even in time of peace, though, there is enough EMP floating around that every digital device must be programmed to check the signals coming in. Before the Control Unit in DIM-I can really be sure it got a signal from the keyboard, therefore, it goes back to the "3" key a few times to make sure the key has actually been pressed.

Having determined that the pulse it heard did in fact come from the "3" key, the Control Unit now proceeds the same way the stationmaster did. First it enters the newly arrived 3 in a logbook and then sends the signal on to its proper destination.

The "logbook" for recording the receipt of a signal from the keyboard is a chain of four transistors constituting a small memory unit known as a "register." The Control Unit sends a signal—a surge of current—to a group of logic gates. This logic circuit, in turn, emits a pattern of electronic pulses to the register, switching the four transistors on and off so that they line up like this:

OFF OFF ON ON

This pattern is the electronic version of the binary number

0 0 1 1

Reading from the right, this means one 1 plus one 2 plus no 4's plus no 8's. Add it all up—1+2+0+0—and 0011 is the binary version of decimal 3.

With that taken care of, the 3 now has to be sent on to its destination—in this case, the calculator's display screen. DIM-I is going to turn the keyboard "input" into display screen "output." That is, it's ready to put that 3 on the screen.

Being at least as dumb as a dull-witted stationmaster, the Control Unit can't even begin without looking up the instructions. It jumps ahead to the memory location that stores the programmed sequence of steps that will display a number in the first digit position of the screen. The calculator will now fetch and execute, in precise rhythm with the ticking of the clock, the instructions to operate the "output decoder" circuitry—the series of logic gates that will transform the binary 0011 that has been stored in the register into the digit 3 on the display screen.

Small digital devices—watches, calculators, games, etc.—generally use one of two standard display techniques. The light-emitting diode (known to the engineers as "LED") system employs a group of small diodes made of a special semiconductor material that work something like a standard light bulb; they give off a bright glow (either red, yellow, or green, depending on the material used) when current flows through. The liquid crystal display ("LCD") is a creamy gray liquid that works the opposite way; the crystal blocks the passage of light, and thus turns black, when current is applied. The result is a black digit floating on a gray background. In either case the display produces the characteristic squared-off digits that have become part and parcel of the modern age, staring out at us everywhere from the digital readouts of clocks and calculators, scoreboards and scales.

Each of the ten digits from 0 to 9, together with few other standard symbols such as a minus sign for negative numbers, can be created in the familiar squared-off style from a rectangular cluster containing seven separate diodes, arranged this way:

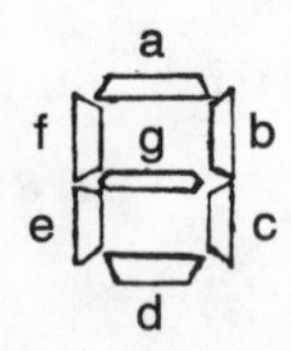

Drawing all the digits using just these pieces turns into a sort of high-tech version of those maddening little puzzles that ask you to form a

"T" out of four weirdly shaped pieces of plastic. In a seven-segment display, some shapes are simpler to make than others. To make a number 8, all seven diode segments would be lighted. A 1 is formed by lighting only segments b and c. A 5 can be made with segments a, f, g, c, and d:

If an eighth diode, shaped like a period, is added to the cluster, this eighth segment can represent a decimal point:

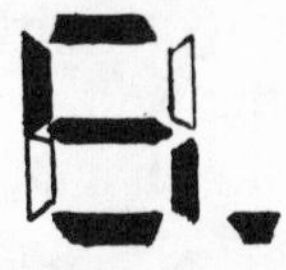

The calculator's output circuit has to solve the puzzle: it has to draw the decimal digit representing whatever number the Control Unit sends along. The output unit consists of a group of logic gates that sends out a pattern of signals to selected segments of the display. In the case at hand, the binary pattern 0011 flows from the Control Unit to the output gates. The gates, in turn, will turn on segments a, b, g, c, d, and h, but not segments e and f. The display, consequently, lights up like this:

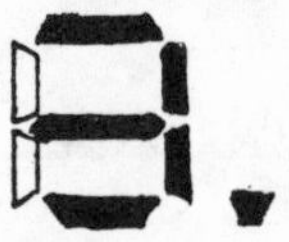

Moving the 3 from keyboard to display screen—and writing it down in a register along the way—has taken hundreds of instruction cycles. If the machine took a whole second to fetch and execute each instruction, we would have waited ten or fifteen minutes for the display to light up. But our calculator is somewhat faster than that. With its clock ticking every hundred-thousandth of a second, DIM-I runs through all the instructions and completes the job in a total time of about .005 second. On the human scale, this is effectively instantaneous; the digit appears to hit the display screen the instant we push the key. The calculator is quicker than the eye.

It's also quicker than the hand. Presumably, the human operator will soon be punching some more keys. But even an unusually nimble human finger will take two or three tenths of a second between keys —virtually eons of time on the microelectronic scale. Thus the Control Unit, even at the unhurried (by microelectronic standards) pace of our 10-microsecond pocket calculator, has more than enough time to check for EMP, write the 3 in a register, send it on to the output circuitry, and return to its familiar old pastime—the idle routine—before the finger can begin to move on.

On its next few hundred queries to the keyboard, consequently, Control will probably find a finger still resting on the "3" key. This manifestation of human torpidity requires a whole new subgroup of the idle instructions to see what the person really has in mind: was he just taking his time about punching that 3, or is he really trying to enter a number like 33 or 33333333?

That settled, the Control Unit can go back to its stationmaster duties: watching the keys, checking each pulse against the timetable, logging it in, and sending the appropriate signals out to various parts of the circuitry. It may switch the pulse onto a wire leading to the main memory unit, to a special memory register, or to the display screen.

And occasionally it sends the signals on to the most important destination in any digital machine. This is a complex nest of logic gates known by various names: "Arithmetic-Logic Unit," "Adder-Subtracter," etc. Whatever it's called, this central arithmetical circuitry is the place where a computer does its computing.

Let's assume, for example, that the person who pushed DIM-I's 3 key was starting to punch in an addition problem: "3+2= ." As we have seen, the first keystroke resulted in an byzantine sequence of operations that left 0011 (binary 3) in a memory register and displayed a 3 on the screen. For the next two keystrokes, the same general sequences are followed again in all their labyrinthine complexity as DIM-I digests the + and the 2.

Things become even more difficult when the pulse comes in from the "=" key. That pulse is the call to action. As soon as control senses that someone has touched the equals key, it jumps to a new set of instructions that dispatches the numbers and the plus sign to the arithmetic unit. The most complicated part of any digital device is the circuitry for that unit. All the switches and all the wiring must be arranged so that no matter what binary pulses are fed in, only the correct answer can come out.

The basic building blocks of the arithmetic circuit are the standard logic gates that Claude Shannon first proposed in 1938. As Shannon showed then, various combinations of switches can be wired together with a few resistors and capacitors to carry out the basic logical operations that George Boole had formulated. The simplest of these is the circuit called a "NOT gate"—a set of switches with an output that is always the direct opposite of the input. That is, if current flows into the NOT gate, no current flows out. When current is not flowing in, current does flow out.

Translating this to binary math, a NOT gate that takes in a 1 sends out a 0, and vice versa. The complete operation of this small, simple circuit can be set forth in a short, simple table:

IN	OUT
1	0
0	1

Because the operation of this circuit is a function of fundamental physical forces, there can be no result other than those shown in the table. Accordingly, such a table is known to the computer engineers as a "truth table."

The standard approach to circuit design is to prepare a truth table that provides the desired result and then, using Shannon's techniques, build a circuit that produces a result to match the table.

It is simple enough, for example, to produce the truth table for the Boolean AND operation. As we saw a few chapters back, the rule of the AND operation—typified by the sleepy commuter who has to get out of bed in the morning if his clock says it is 7:00 A.M. AND his calendar says it is a workday—is that the result is yes only if condition A AND condition B are both yes. Stated in binary terms, the output is 1 only if input A AND input B are both 1. Set down in a truth table, the AND operation looks like this:

A	B	Result
0	0	0
0	1	0
1	0	0
1	1	1

It is possible for these things to become a whole lot more complicated. An AND operation, for example, could have three or four or eight different inputs—the output is 1 only if input A AND B AND C, etc., are all 1—leading to a considerably longer truth table.

Further, different gates can be chained one after another to achieve different logical results. In a common instance, a NOT gate is wired to the output of an AND gate, so that the result of the AND operation is directly reversed. This combination of NOT plus AND creates a new gate called a NAND gate, which has become, for various technical reasons, the most popular building block of all for many computer designers.

To build an addition circuit, then, that can add two digits and come up with the truthful answer every time, the computer designer first draws up the truth table for a two-input addition problem. This table encompasses all the rules for binary addition:

1st digit	2nd digit	Sum
0	0	0
0	1	1
1	0	1
1	1	10

(The 10, which is read "one-zero," is the binary version of 2.)

The engineers have perfected several different circuits that will produce precisely this binary result. DIM-I employs the simplest, a circuit constructed from about a dozen standard gates. The electronic pulses flow into the gates in patterns representing the two binary digits to be added. What comes out is a new pattern of pulses representing the sum of those two digits—and a carry digit, if necessary, to be carried to the next column.

When the problem "3+2" is punched into the keyboard of DIM-I, the Control Unit stores the two numbers in binary form in a pair of registers and then feeds them into the addition circuit, one column at a time:

$$\begin{array}{r} 0011 \\ +0010 \\ \hline \end{array}$$

starting with the right-hand column, each pair of digits enters the logic gates. Because of the way the gates are wired, there is only one pattern of pulses that can possibly emerge: "0101."

Now the Control Unit gets back into the act. It sends a signal to the output circuitry: "Get ready to display the answer." The binary pulses flow in; current flows out along selected wires leading to the display. The chosen segments light up, and the binary answer to the problem, 0101, is translated on the screen to its decimal equivalent:

Ta da! In the most roundabout conceivable fashion, DIM-I has actually solved a problem. In the minute step-by-step manner of the microprocessor, though, this seemingly simple task has become a full-scale production. The solution of 3+2, a problem any human could work in a single mental flash, had to be broken down for the calculator into thousands and thousands of individual instructions that led the machine by the nose through its paces. It is a process that only a mindless mechanism could love. It is process that only a powerful intelligence could have invented. But then, digital mechanisms were invented by the most powerful computer on earth—the human mind.

10

"It's so simple"

A SPECTACULARLY ROCOCO official document showed up in Jack Kilby's mailbox in Dallas on the morning of June 27, 1974. Framed in a silken strip of bright red ribbon, affixed with a gleaming gold foil seal, engraved in a florid cursive script, and illustrated with an ornate etching that depicted, among much else, the American eagle, the stars and stripes, assorted machinery, and the headquarters building of the U.S. Department of Commerce, the document poured forth its message in a rich flood of legalistic prose:

> TO ALL TO WHOM THESE PRESENTS SHALL COME: WHEREAS Jack S. Kilby; Jerry D. Merryman; James H. Van Tassel, all of Dallas, Texas . . . presented to the COMMISSIONER OF PATENTS a petition praying for the grant of LETTERS PATENT for an alleged new and useful invention the title and a description of which are contained in the specification of which a copy is hereunto annexed and made part hereof . . . and WHEREAS upon due examination made the said Claimants are adjudged to be justly entitled to a patent under the law: Now therefore these LETTERS PATENT are to grant . . . for the term of SEVENTEEN years from the date of this grant.

The specification thereunto annexed and made part thereof was, in fact, the patent application that Texas Instruments had sent off years earlier when Kilby and his "Cal Tech" team had successfully developed the first pocket calculator. With the formal receipt of the patent, T.I. could now replace the words "Patent pending" on its pocket calculators with the official patent number: U.S. Pat. 3,819,921—a legend that is still being molded today into the plastic cases of calculators turned out by Texas Instruments and other firms it has licensed to copy the basic Kilby architecture.

In sharp contrast to the long legal struggle over the integrated circuit, there had been no contest about the patent for the handheld calculator. Everyone recognized that Kilby, Merryman, and Van Tassel were entitled to the honor. By the time the patent was issued, however, the honor of the thing was just about all it stood for. By the middle of 1974, it was clear that Texas Instruments and all other American manufacturers were losing control of the calculator business. This revolutionary product—conceived, born, and bred in Texas, U.S.A.—was becoming a Japanese preserve.

Given the Japanese dominance of other facets of the consumer electronics business, it was not altogether surprising that firms like Casio and Panasonic had caught and passed their American competitors. But it was still a bitter irony. By succeeding with the pocket calculator, the Japanese electronics firms had added a sour ending to one of the happier chapters in contemporary American business history.

Since its first appearance in the 1930s, the office calculator had been recognized worldwide as an American product. Despite impressive competition from European firms, U.S. manufacturers dominated the market. Then, in the 1960s, when a "calculator" was still a hefty, expensive, desk-top apparatus, Japanese manufacturers had moved into the U.S. market and simply overwhelmed the traditional American office-machine firms. Japan's market dominance increased steadily, and the situation was bleak for the American side until the day in 1971 when Jack Kilby's first handheld calculator went on the market. The application of microelectronics to the calculator business completely turned the tables. The chip-based calculator—smaller, cheaper, and more efficient than any Japanese product—put American labels back on top.

This reversal was so dramatic that the Commerce Department issued a detailed report on the American triumph: "With innovative products and aggressive pricing policies backed by high volume production efficiencies, U.S. firms rapidly regained control of the U.S. calculator market from their chief competitors, the Japanese." The report went on to say that this turnabout was "an excellent illustration of the way a free and competitive industrial system can provide benefits to the consumer," and a model that other U.S. industries could emulate in their struggles against foreign competition.

Unfortunately, the U.S. firms were not the only ones in the calculator business who knew how to practice aggressive pricing and production efficiencies. By the time Jack Kilby received his Letters Patent, Japanese manufacturers had obtained the information they needed to make handheld calculators of their own. In part, they developed their own calculator know-how; mainly, though, they borrowed American techniques, either buying or appropriating patented processes from U.S. firms. (T.I. sued Casio in 1982 for infringement of the Kilby calculator patent; the case was settled with Casio paying an unspecified sum.)

In any case, the competition from Japan was so intense and so successful that, two years after its original calculator report, the Commerce Department had no choice but to issue an unhappy update: "Since 1974, the situation has once again been reversed. . . . The departure of U.S. firms from the calculator industry continues. . . . Japanese-origin calculator imports have again been a major factor in this downward trend." When that was written, in 1977, the Japanese made about 45 percent of the calculators sold in the United States. By the early 1980s the Japanese firms' share of the U.S. market, in dollar value, was above 70 percent.

As the Commerce Department noted, the pattern of conquest in calculators was familiar to anyone who had watched Japanese firms make equally successful forays into U.S. markets ranging from automobiles to zippers. The Japanese had first moved into the low-priced segment of the business, turning out simple, "plain vanilla" calculators that offered the consumer greater variety, equal or better reliability, and lower prices than comparable U.S.–made brands. Having established a beachhead at the low end of the market, the invaders moved to broaden their territory. "All they did was take our product and add a bunch of bells and whistles," a U.S. manufacturer complained. But the bells and whistles—not to mention the calorie counters, currency converters, perpetual calendars, astrology tables, biorhythm charts, and assorted other add-ons dreamed up by the marketing men in Tokyo—sold calculators by the millions. In the third stage of their assault, the Japanese set their sights on more sophisticated, higher-priced models that provided the highest profit margins. "This pattern of market activity," the Commerce Department's second report said, "has been observed for other products, such as television sets."

The comparison to television sets had to be a frightening one for anybody in the U.S. electronics industry. Television is the paradigm case of international competition in modern electronics, but it is hardly a pleasant paradigm for American industrialists to ponder. Just about every significant advance in the development of television has come from the United States. Yet the Japanese have trumped the American genius for innovation with their own skills at production and marketing.

The ancestry of the television set can be traced to the work of several late-nineteenth-century European physicists who developed and perfected the cathode ray tube. The modern TV is a direct descendant of this device, a lineage reflected in the slang terms "tube" and "boob tube." The cathode ray tube, originally built strictly as laboratory apparatus, was a long, closed glass tube with a mechanism inside that shot a "cathode ray"—the mysterious beam of energy that J. J. Thomson determined to be a stream of electrons—down the length of the tube. At the opposite end of the tube the glass was coated with phosphor. An electron shot from the cathode gun raced down the tube and struck the phosphorescent glass at the far end. Precisely where the electron hit, the phosphor would glow with a bright point of light.

Peering at an interesting pattern of glowing dots on a cathode ray tube in 1908, the British physicist A. A. Campbell-Swinton hit upon the fundamental idea of television: the dots of light on the glass tube could form a pointillistic picture—the television image. (If you put down this book for a moment and look closely, very closely, at a television screen, you will see that the picture on the glass is actually composed of thousands of individual points of light, à la Seurat.) Campbell-Swinton named his idea "Distant Electrical Vision"; the term "television" was apparently coined by the French. As every schoolchild in the Highlands knows, the Scotsman John Logie Baird pursued the idea with some success. But the real invention of television came about in the 1920s in a pair of American laboratories run by a pair of competing inventors who battled furiously to be first to perfect the concept.

In one corner was Vladimir Zworykin, an officer in the Tsar's Army who had fled Russia during the 1917 revolution and eventually landed in David Sarnoff's laboratory at RCA in New York. In the other was Philo T. Farnsworth, a fiercely independent farm boy from Idaho who would not sign on with any of the major electronics firms and was forced to scrounge constantly for money to support his research. (After years of work, and just on the verge of final success, Farnsworth retired back to the farm at the age of thirty-four—a classic burnt-out case.)

Spurred on by the whip of competition with each other, Zworykin and Farnsworth rapidly surpassed the efforts of Baird and other Europeans, and by 1929 they had developed, between them, a camera that could convert a visual image into electronic signals and a picture tube that could translate these signals back into the visual image. The heart of the Zworykin/Farnsworth TV—and of almost all television sets today—is a cathode that fires a moving beam of electrons toward the phosphorescent screen. The electron gun sprays a picture on the screen the way an aerosol paint can sprays a picture on the subway wall. The graffiti, of course, is on the wall for good; the television tube, in contrast, sprays a new image on the screen sixty times each second, creating the illusion of a moving picture.

Serious commercial development of television was delayed until after World War II, when the U.S. government finally got around to issuing technical regulations for frequencies, band widths, and the like, so that all stations could broadcast the same standard signal. In the first year after the war, 6,000 television sets were sold in the United States. In 1948, Americans bought 1 million sets; two years later, 7 million. The boom was on. Because American technology set the pace, the U.S. standards were adopted, with minor variations, around the world.

As their sales skyrocketed, U.S. television manufacturers developed a marketing structure not unlike that of another booming American business, the automobile industry. The major television makers—Admiral, DuMont, Sylvania, etc.—sold their products through franchise dealerships that generally handled only one brand. The dealers provided factory-authorized repair, an important function at a time when the family TV was likely to be in the shop as often as the family car.

And they dutifully pushed the large console models that brought manufacturers the largest profit margins.

This retail structure made it doubly hard for new firms—particularly overseas manufacturers—to move into the U.S. market. There was no way for an outsider to break into the dealership system, which was controlled by the big domestic firms. Further, without a network of parts depots and factory-trained repairmen, how could an overseas competitor keep his sets in working order?

These problems persuaded the major European electronics firms to stay home. But the Japanese manufacturers, prodded by government officials eager to increase exports, took a long look at the United States in the early 1960s and decided to give it a try.

On technical matters, the Japanese initially copied what others had done. The firms purchased the rights to more than 400 patents from U.S. and European firms and generally waited for others to make the important breakthroughs. In production and marketing, however, Japan did the innovating.

The Japanese television firms approached the U.S. market with a wider variety of models than many domestic makers offered, including a line of smaller table-model sets and the first portables. By targeting the low-price segment of the market, they were able to sell their sets through the fast-growing discount chains, circumventing the traditional dealer network. To deal with the lack of trained repairmen, the Japanese set out to eliminate the need for repair. They honed a highly disciplined labor force and introduced extensive quality-control mechanisms in their factories, so that Japanese television sets would work reliably without constant maintenance.

The results were dramatic. In 1961 the six Japanese manufacturers combined sold less than one-half of one percent of the black-and-white television sets purchased in the United States. They moved up to 2 percent of the market in 1962 and 13 percent in 1965; thereafter, the percentage kept going up, year after year.

The American manufacturers shrugged off this challenge. By the mid-sixties, they were concentrating on an even hotter new product—color television, in which U.S. technology led the world by a large margin. But the color TV market turned into a rerun of the monochrome story. The Japanese aggressively purchased patent licenses in the West. They entered the fray a few years behind the American firms but made up for the late start with their familiar combination of extensive variety, low prices, and high quality. By 1976, Japan was the world's leading producer of color television sets and had 35 percent of the U.S. market.

At this juncture, some U.S. firms simply gave up. Admiral quit the business. Motorola sold its television operation (the "Quasar" line) to Matsushita. Others fought back, both in the market and in Washington. U.S. manufacturers charged, and eventually proved, that some Japanese competitors were "dumping" TV sets in the U.S. at predatory prices; under the law, this could have resulted in multimillion-dollar fines for the Japanese firms, but diplomatic considerations prevented that result. Instead, the governments of Japan and the U.S.A. in 1977 negotiated an "Orderly Marketing Agreement"—the diplomatic term for an import quota. Those yearly agreements preserved almost 80 percent of the U.S. market for domestic makers (the term "domestic," in this context, includes Japanese firms building sets in the United States). In the end, it was diplomacy, not technology or business acumen, that saved the United States a piece of the action in a market it had pioneered.

The television saga would be an unsettling one from the American point of view even if it were the sole instance of such an abrupt industrial defeat. But in consumer electronics, as in automobiles and steel, the history just related has been the rule, not the exception. The pattern regularly recurs: despite unchallenged technological leadership, American firms have been overtaken by Japanese manufacturers offering a broadly appealing range of products that combine premium quality with competitive prices.

The Commerce Department, displaying the universal bureaucratic yen for numbers and tables, issued a study in 1980 that reduced the whole sorry state of affairs to a single chart:

Import Penetration—Consumer Electronics

Product	Imports as % of U.S. sales
Videotape equipment	100%
Household radios	100
CB radios	90
TV—monochrome	85
Electronic watches	68
Hi-fi, stereo components	64
Tape recorders (audio)	35
Microwave ovens	25
TV—color	18

In each case, the product was conceived and developed in the United States; in each case, the Japanese (sometimes followed by producers from other nations, such as Taiwan and South Korea) borrowed the technology and then swept into the American market.

The American semiconductor industry, during the first booming decade after Jack Kilby and Bob Noyce hit upon The Monolithic Idea, could safely look upon these developments in consumer electronics as irrelevant. In chips, the American pioneers had an enormous technological lead over foreign competitors everywhere, European or Asian. And however daunting the Japanese might be in consumer electronic goods, the chip was a different animal. The Japanese had proven adept at moving into relatively stable technological fields, where the design of the product was settled, and producing large volumes of goods at reasonable prices. But the integrated circuit business, in its first quarter-century, was never stable. Improvements came so rapidly that there was no such thing, really, as a settled design; prices fell so rapidly along the learning curve that undercutting from foreign competitors did not loom as a serious threat.

Still, there were signs during the 1960s and early '70s that the Japanese had a collective eye on the semiconductor business. The big Japanese electronics firms—outfits like Fujitsu, Nippon Electric Company (NEC), Hitachi, and Sony—began buying licenses for every U.S. semiconductor patent they could find; between 1964 and 1970, royalty payments from Japan on U.S. semiconductor patents rose by a factor of ten, from $2.6 million to $25 million per year. For some U.S. firms, particularly Fairchild, this became an important source of no-risk income. "American firms have generally been very cooperative," a Brookings Institution economist, John Tilton, reported in 1972. "With few exceptions, they have been willing to license Japanese as well as other foreign firms and aid them in assimilating new semiconductor technologies, even though in the process they are helping establish potential rivals."

There was only one major U.S. firm that refused to cooperate—Texas Instruments. T.I. spurned royalty payments and set a more ambitious price on its semiconductor patents: no Japanese firm could use them unless the Japanese government permitted T.I. to set up manufacturing operations in Japan. This demand posed a serious dilemma in Tokyo. The Japanese were determined to exclude all foreign competition from Japan. But in order to produce integrated circuits, the Japanese firms would need access to both the Noyce patent from Fairchild and the Kilby patent from Texas Instruments. Fairchild was willing to sell its technology, for a royalty of 4.5 cents on every dollar the Japanese makers earned on chips. But the Texans hung tough. In 1968, after several years of offers and counteroffers, the Dallas firm was finally permitted—in return for a license to the Kilby patent—to open a Japanese plant. For more than a decade thereafter, T.I. was the only U.S. semiconductor producer with any significant sales in Japan.

Whether or not the American firms were aware of it, both the vigorous pursuit of U.S. patent rights and the exclusion of U.S. competition were part of a grand strategy devised by the Japanese government to win that nation global preeminence in chips. The semiconductor business was one of several high-tech industries targeted by the Japanese Ministry of International Trade and Industry (MITI) for intensive development. In its ten-year plan, or "Vision," for the 1980s, MITI concluded that microelectronics was perfect for the island nation because it required large quantities of human resources, such as advanced engineering and diligent workers, which Japan has in abundance, but only small amounts of energy and natural resources, which Japan lacks.

Officially, MITI's decisions about what's good for Japanese industry constitute nothing more than "administrative guidance," which companies are legally free to ignore. In the 1950s, for example, someone at MITI had the brilliant insight that Europeans and Americans would never shell out their money for cars bearing names like "Honda" or "Toyota." MITI issued "guidance" telling a group of auto makers to get together and design a single "people's car" that would represent the entire Japanese industry around the world. Honda, Toyota, Nissan, etc., rejected this idea, with stunning worldwide results. When MITI worked with its electronics industry in the sixties to plan Japan's foray into the low-priced end of the U.S. television market, one manufacturer, Sony, spurned the consensus view and found a lucrative niche of its own as a prestige upper-bracket label.

As a rule, though, MITI's policies tend to become industrial practice. This certainly seems to have happened in semiconductors. Just as MITI planned it, Japanese electronics firms acquired the U.S. technology required to get started in the manufacture of integrated circuits in the mid-sixties. Just as MITI planned it, these firms were able to rely on sales to domestic Japanese computer and telecommunications firms—free of competition from any U.S. firm except Texas Instruments—to provide the financial cushion necessary for an assault on the world market for chips. And just as MITI proposed, the five largest electronics firms banded together in the early 1970s for a cooperative research endeavor—funded partly by the government and partly by the companies—to develop manufacturing techniques for very-large-scale integrated (VLSI) chips.

For all this effort, however, the Japanese electronics industry still lagged far behind American firms in most areas of microelectronics. The one product in which the Japanese took any significant market share was Random Access Memory, or RAM, chips. The memory circuit, pioneered by Robert Noyce and Gordon Moore at Intel in 1968, was nicely suited to the Japanese strengths: a simple and relatively stable product that was consumed in huge quantities by computer manufacturers around the world. Even in the RAM market, though, U.S. dominance was unchallenged until the middle of the 1970s. Then the American industry gave Japan its chance.

During the prolonged recession following the 1973 oil embargo, American semiconductor firms reacted the way American manufacturers normally react to recessions: they laid off workers, closed plants, and generally hunkered down to await an upturn. In 1976, when the economy came roaring back, there was an enormous burst of demand from computer firms for what was then the most advanced memory chip: the 16K RAM, capable of storing some 16,000 bits of information. The U.S. firms could not rebuild fast enough to meet the need. Their customers went shopping for an alternate source of RAM chips—and found it in Japan.

Following standard Japanese industrial practice, the big Japanese electronics firms had maintained their work force and their production capacity during the recession. They were, accordingly, in the catbird seat when the market for 16K RAMs took off. Silicon Valley's inability to meet demand gave Japan a golden opportunity to show the world what it could do, and the Japanese firms leaped at the chance. They began dispatching high-quality, competitively priced chips around the world. By 1980 the Japanese had 42 percent of the world market.

More important, the Japanese firms had put themselves in perfect position to compete when the next generation of memory chips, the 64K RAM, was developed in the late 1970s. This time, Japanese firms had no technological lag to worry about, and this time they had established markets everywhere for their memory chips. After two years of nip-and-tuck competition, Japanese firms finally eclipsed the Americans. By 1983 they controlled well over half the world market for 64K RAMs.

The Japanese success in this one product line prompted all manner of distressed breast-beating in the United States—far more than the situation actually warranted. The RAM memory chips constitute a high-volume part of the semiconductor market but not a particularly important or remunerative one. In dollar terms RAM chips probably represent somewhere between 5 and 10 percent of annual semiconductor sales. Americans have maintained supremacy in virtually every other type of integrated circuit. For Silicon Valley to worry about Japanese sales of memory chips is like General Motors losing sleep over a smaller company that has done well in the spark plug business.

But many Americans were worried. The Japanese inroads in chips, a congressional committee reported, "indicate the potential for an irreversible loss of world leadership by U.S. firms in the innovation and diffusion of semiconductor technology." The Silicon Valley firms were sufficiently alarmed to form a trade group, the Semiconductor Industry Association, specifically to fight off the foreign challenge. "The television people woke up when the Japanese had 20 percent of the market, and went to the government when the Japanese had 40 percent," the group's executive director told *The New York Times*. "That's a little late."

The Semiconductor Industry Association began turning out a steady flow of studies and brochures and advertisements, along with petitions to various government commissions and agencies, asserting that the competition from Japan was unfair. The SIA cited MITI's determination to keep U.S. companies out of the Japanese market. It complained about Japanese government contributions—something over $100 million—to the very-large-scale integration research and development venture. It pointed out that the Japanese firms, partly because of that nation's industrial structure and partly because of support from MITI, received loans from Japanese banks at much lower interest rates than any electronics firms could hope for in the U.S.

"I just don't want to pretend I'm in a fair fight. I'm not," wrote W. J. Sanders, chairman of Advanced Micro Devices, in a statement that crystallized the SIA position. "Do you know how the Japanese got the dynamic RAM business? They bought it. (If I had their deal, I'd have bought it too.) They pay 6 percent, maybe 7 percent, for capital. I pay 18 percent on a good day. . . . They start every product development cycle with hundreds of millions of dollars of free R&D every year, paid for by their government. Good for them. But then their parts arrive here in a flood."

There was, however, another point of view—a view held, among other places, at Texas Instruments, which never found reason to become a member of the Semiconductor Industry Association. The dissenters pointed out that despite Japan's efforts to exclude foreign competitors, American firms have always had a larger share of the Japanese semiconductor market than the Japanese have gained in the U.S. They argued that the American semiconductor industry, launched on a wave of government financing and now receiving tens of millions of dollars annually from the Pentagon for research and development, is hardly in a position to carp at MITI's grants to Japanese firms.

The SIA's American critics also recalled that the Silicon Valley firms, by selling their patents and by cutting capacity just before the 1976 boom, were themselves partly responsible for the Japanese success. "Those fellows on the West Coast sort of have schizophrenia," Fred Bucy, the president of Texas Instruments, said after the SIA was founded. "They had the same leverage as we did. . . . But they were very shortsighted in the way they handled the patent situation."

There was one other factor, too, fueling the Japanese success in RAM memory chips, but it was something that American trade groups were not eager to talk about. It was a familiar factor to Americans who had observed the Japanese success in television, in cameras, in automobiles—the quality factor.

In the middle 1970s, when the Japanese electronics firms first managed to break into the American semiconductor market, U.S. companies buying imported memory circuits for installation in their computers began to notice something interesting: Japanese chips were better.

There was no discernible difference in performance, because all memory chips are built to work the same way. But there was a marked difference in reliability. Chips made in Japan were less likely to break down than the American product.

At first, the Japanese edge in quality was the dirty little secret of Silicon Valley. Hardly anybody talked about it, and when the subject did come up, American manufacturers heatedly denied that Japanese firms were turning out more reliable chips. That changed one morning in March of 1980 when an American computer executive named Richard W. Anderson stood up at an industry meeting in Washington, D.C., and delivered a paper that has come to be known as "The Anderson Bombshell."

Anderson was a division manager at Hewlett-Packard, a giant California-based manufacturer of electronics instruments and computers that is one of the world's biggest consumers of integrated circuits. It was he who had decided, rather reluctantly, to start buying Japanese memory chips for Hewlett-Packard computers, and at the Washington meeting he told his story.

"We first introduced semiconductor memory in our computers in 1974," Anderson said. "We got all our memory from United States suppliers. Then in 1977 the 16K, or 16 thousand-bit, RAM began to make its appearance. The first introductions that I was familiar with came from U.S. suppliers, and we hurried to implement this design into our product line. . . .

"However, some months after introduction, the U.S. suppliers that we had been working with found themselves unable to meet our quantity demands, either due to yield or capacity problems, and this left us between the proverbial rock and a hard spot. So, after much anguish, we decided to talk to a Japanese company who had been calling on us telling us of their memory for some time. And I would like to state at the outset we took a very cautious approach because we remembered well the impressions from post–World War II Japanese products; namely, that they were cheap, low cost, and low quality. And so our engineers went through a very rigorous qualification program; and we were pleasantly surprised to find they qualified."

Anderson went on to say that, over time, he bought more and more chips from the Japanese firm. Although the fact was not immediately obvious, he said, Hewlett-Packard gradually began to realize that there was a significant difference in the Japanese memory circuits. "We had fewer failures in incoming inspection; we had fewer failures during production cycle; we saw fewer failures of products in customers' hands. . . . Not only was the quality good, but [it] was actually superior to what had been our experience with the domestic suppliers.

"Then came 1979," Anderson went on, "and a real market crunch hit the memory suppliers, particularly the U.S. manufacturers . . . and we found ourselves in short supply. So we went back to Japan and qualified two more Japanese suppliers for the product line that I'm responsible for. And again the same experience: excellent quality." Eventually, Anderson added, Hewlett-Packard compiled performance records on some 300,000 memory chips, of which half came from the Japanese suppliers and half from American makers. The final standings showed that all three Japanese firms were delivering higher quality goods than the best American manufacturer. "So that's a remarkable, and I would think to American suppliers, perhaps a frightening set of statistics," Anderson said.

Frightening it certainly was. The message that foreign competitors were outperforming the U.S. in semiconductors, the symbol of American technical preeminence, was a slap in the face that could not be ignored. The Anderson Bombshell, widely reported and corroborated by some other computer firms, made it impossible for American firms to deny any longer that there was a quality difference. Instead, they set out to learn how the Japanese had attained it. "U.S. Microelectronics Firms Study Japan for Secrets of Quality and Productivity," read a headline in *The Wall Street Journal* in 1981. The spate of books that appeared in the early 1980s extolling Japanese management practices became required reading in the semiconductor business. The American Electronics Association did a booming trade in seminars on Japanese quality control. Companies dispatched fact-finding teams to Tokyo to uncover the Japanese quality secret.

There was a strong streak of irony in this reaction. As the Japanese knew perfectly well, their quality mechanisms were hardly secret—particularly not to Americans. After all, Japan had learned virtually all it knew about quality control from an American. The teacher was a federal bureaucrat, an obscure fellow, unknown in his homeland but recognized and respected throughout the Japanese islands. He is a son of the prairie named W. Edwards Deming.

W. Edwards Deming was born in Sioux City, Iowa, in 1900. Growing up on the prairie, young Deming, like many intellectually gifted youngsters, displayed talent both in mathematics and music. Eventually, the lure of math and science prevailed, and as an undergraduate at the University of Wyoming, Deming concentrated on math, physics, and engineering. He went on to graduate work at the Colorado School of Mines and at Yale, where he took a Ph.D. in physics in 1928. Dr. Deming then began a long career in federal service, first as a scientist in the U.S. Department of Agriculture's Bureau of Soils and later, when statistics became his dominant professional interest, at the Census Bureau. Retired now, he runs an extraordinarily vigorous lecturing and consulting operation that is headquartered in the cluttered basement of his Washington home but keeps him on the go, all over the world, about three hundred days per year.

Deming is a tall, crusty, imposing gentleman with a fluffy white crew-cut, spectacles, and a jutting nose that gives him the look of a tortoise poking around outside the shell. He has always been a prodigious worker; he has, in fact, developed a method of squeezing two workdays out of one: he works until mid-afternoon, naps until eight-thirty at night, and then rises for several more hours of business. He finds time, naturally, for music—sessions at the piano with his wife, Lola Shupe Deming, and their children and grandchildren. He composes liturgical music; the Deming *oeuvre* includes two masses and a number of canticles. Otherwise, he is all business. I asked him once—he was eighty-three at the time—if he ever takes a vacation. "Well, I'm not doing anything for the next few hours," came the gruff reply.

A nine-page "List of principal papers" that Deming provides his consulting clients reflects the changing focus of his interests over the years. The first dozen or so of his publications—many signed jointly by W. Edwards Deming and Lola Shupe—deal with physics and chemistry: "Equipotential surface electrons as an explanation of the packing effect"; "Note on the heat capacity of gases at low pressure." But midway through the 1930s the titles begin to reveal a developing interest in probability and statistics: "On the frequency interpretation of inverse probability"; "On a least squares adjustment of a sampled frequency table when the expected marginal totals are known"; "On the elimination of unknown ages in the 1940 population census."

This shift in Deming's scholarly output resulted from his growing appreciation for the enormous power of statistics to explain precisely what was going on in any repetitive process—including mass production of commercial goods. Deming learned that the statistician could act as a skilled detective when things went wrong in an industrial operation, searching his data to pinpoint the problem and eliminate it. He put this idea on paper in a seminal monograph, written in 1934 and updated thereafter, titled "On the Statistical Theory of Errors."

The "Theory of Errors" was Deming's explication of the ideas of Dr. Walter Shewhart, the American mathematician who first applied statistical methods to the control of factory operations. Deming set out to be the evangelist of the Shewhart gospel, which held that careful statistical records were the essential key to consistent quality in manufacturing.

In the lectures he delivers around the world today, and in his basic text, *Quality, Productivity, and Competitive Position,* Deming is still preaching the lessons he set forth fifty years ago in the Theory of Errors. He stresses that quality and productivity result from diligent observance of certain fundamental rules—he has gathered them into a table of "14 points"—all of which can be summarized in a single, golden rule that should govern all work: "Do it right the first time."

Reduced to those six words, the Deming message seems obvious—a point the professor never fails to drive home. "It's so simple," he says in his lectures. "It's so obvious." But to carry out the obvious requires detailed effort. To achieve consistent quality, those involved in any operation, from manufacturing semiconductors to managing a baseball team, must maintain "statistical control"—that is, careful, regular measurement of all aspects of the job. "By describing statistically exactly what is done," Deming says, "the method locates your problems and leads to innovations that solve them."

One time, for example, Deming was retained by a shoe company that had run into costly but inexplicable manufacturing delays. Management was baffled; the same work force in the same factory had suddenly fallen far behind its normal production rate. Poring over time cards, maintenance charts, purchasing records and the like, Deming found that someone had recently ordered a new brand of thread. Now it was all so obvious. The new, cheaper thread kept breaking, forcing workers to stop and rethread their sewing machines time and again. "To save 15 cents per spool they were losing $150 per hour rethreading the stuff," he explains. Better thread was purchased and the problem disappeared.

Deming asserts pointedly that statistical control is the only sensible means of quality control. "All these companies are running advertisements on the TV about their rigorous inspectors," he snorts. "Inspection is too late, don't you see? It's so obvious. By the time the product gets to an inspector, the quality, either good or bad, is already in. Do you want to burn the toast and scrape it, burn the toast and scrape it—or do you want to make the toast right before it gets to your inspectors?"

The right way to make toast, under the Deming rules, would involve careful observation of every element of the process—the quality of the bread in the market, the wiring and the timer in the toaster, the methods used by the person charged with making toast. If all these variables were tracked on statistical control charts, a manager could see just what had happened when any piece was burned. The problem could be quickly corrected. The control process would be more efficient and better for workers' morale than inspecting and scraping every piece of toast.

Moreover—and this is the Deming route to enhanced productivity—statistical control is cheaper. It will always cost less to make one piece of toast right than to burn one piece, inspect it, reject it, and *then* make one right. "The total cost to produce and dispose of a defective item exceeds the cost to produce a good one," the professor's text says. The time and money previously spent to inspect and scrape burnt toast can be directed to a more productive goal—making more toast—once a predictable level of toast-making quality is achieved.

In the past two decades American industry has learned the hard way that Deming must be heeded. In the 1930s, though, things were sharply different; when W. E. Deming talked, nobody listened. There was a brief spurt of interest in statistical quality control at the start of World War II, when quality of production became a matter of national survival. Engineers from munitions plants were brought in to learn from Professor Deming, and the War Department gave courses in factories. "Brilliant applications burned, sputtered, fizzled, and died out," Deming wrote later. "Quality control departments sprouted. They plotted charts, looked at them, and filed them. They took quality control away from everybody else, which was of course entirely wrong, as quality control is everyone's job."

Deming is not a man to give up easily, but the wartime experience led him to focus on other applications of statistics, including demographics. In the global reconstruction at the end of the war, he became a sort of roving ambassador of demographics. He was an Official Observer at the Greek elections in 1946 and from there went to Delhi to advise on the Indian census. Eventually he was summoned to Japan by Gen. Douglas MacArthur to assist in various population and housing studies. And there, at long last, Deming found an audience that cared about quality control.

A Japanese statistician, Dr. E. E. Nishibori, who was familiar with the "Theory of Errors," discovered more or less by accident that the author of that insightful paper was in Japan. Nishibori tracked Deming down: would the American be interested in speaking at a quality control workshop for the Union of Japanese Science and Engineering? He would. He did. Another workshop followed, and another. Someone else arranged for Deming to speak in Tokyo at the Industry Club of Japan, a business roundtable whose membership included senior management of every major Japanese manufacturing concern. In that speech, on July 26, 1950, Deming declared, to the astonishment of his audience, that Japanese quality would soon be the best in the world.

"I predicted," Deming recalled with great relish three decades later, "that Japanese manufacturers would come to dominate world markets and have their competitors crying for protection. At that time I was the only person in the world who believed that. . . . But it was so simple. You could see that this society was receptive to the ideas that are necessary for quality and productivity."

That speech turned out to be decisive. "Once you convince senior management that this will make a difference, the hardest job is done," Deming says. The Japanese were so convinced that they invited Deming back to Tokyo year after year to spread his gospel. The Union of Science and Engineering distributed millions of copies of his books and pamphlets. Tens of thousands of Japanese factory workers, engineers, and executives studied his teachings at regular classes (they still do so today). These diligent students put his ideas into practice, with consequences that shook the world.

When Deming arrived in the shattered remnants of postwar Tokyo, Japanese goods were everywhere considered cheap, shoddy products made from cheap, shoddy materials. Today the once-proud American automobile industry buys television advertising to boast that its products are made the Japanese way. Today the label "Made in Japan" represents the world standard of quality for products ranging from steel, automobiles and heavy machinery to cameras, scientific instruments and consumer electronic gear.

There is no single explanation for Japan's postwar economic miracle, but the Japanese, at least, give a good deal of the credit to their American teacher. Deming is the only living American who holds the Second Order of the Sacred Treasure, Japan's premier imperial honor. His name and profile adorn the *Demingu Sho* (The Deming Prize), an annual industrial award carrying the stature of the American Pulitzer prizes which is awarded, Oscar-style, in an annual presentation telecast live to the entire nation. The name *Demingu* has become virtually a household word.

And now, fifty years late, the name Deming has come to have meaning in other countries as well—even in the United States. The Japanese conquest has led American management to pay tardy heed to the native-born Cassandra; if it were not so, Deming would not have to squeeze two workdays into one.

Sometime between his seventy-fifth and eightieth birthdays, Deming emerged as the guru of a what is being called the "Third Wave" of industrialization, following the First Wave at the start of the Industrial Revolution and the Second Wave which swept in eighty years ago with techniques of mass production. The Third Wave, suited to a world with higher standards and a clearer recognition of limits, focuses on the most efficient possible uses of resources to produce goods of predictable reliability and quality. The worldwide economic struggle among manufacturing nations today is, to a large extent, the struggle to stay atop that Third Wave.

With characteristic asperity, Deming points out that he saw the new wave coming long ago. "The way to compete in an international economy is to promote efficiency, to produce better quality than the other people are producing," he says. "It's so simple. It's so obvious."

The point was somewhat less than obvious to the American semiconductor industry until the startling Japanese success in memory chips brought the message home quite clearly. The Americans took it to heart. The semiconductor firms made quality control—not just inspection but genuine control—a high-priority challenge. It made a difference. In an interview two years after he released his bombshell, Richard Anderson, the Hewlett-Packard executive, said his firm had found marked improvement in the quality of memory chips coming from American suppliers; in fact, he said, the Americans had matched the Japanese on quality standards. This change helped the U.S. firms take back a small share of the world market for 64K Random Access Memory chips from the Japanese. More important, the new emphasis on quality meant the United States would be able to battle the Japanese on equal terms in the global struggle to market the 256K RAM chip, the state-of-the-art memory circuit during the middle 1980s.

"U.S. manufacturers, until the advent of Japanese competition over quality, had made a tacit decision that fast, volume output with component testing to cover imperfections in the manufacturing process was more important than high quality," reported a study published by Congress in 1982. "The Japanese instead concentrated on perfecting their production process to deliver higher quality devices. As U.S. firms retool and expand capacity, they have apparently been 'tweaking' their production process to deliver higher quality devices. It may well be, then, that the Japanese ability to use quality as a penetration strategy will not carry over to the next round of competition."

Despite this auspicious news, the Semiconductor Industry Association, ever wary, continued to badger the government for help to stave off the Asian invaders. The industry did not ask for direct tariff protection, because that course seemed likely to bring about countertariffs overseas. Instead, the SIA's shopping list in Washington included federal tax preferences, government support of education in electronic engineering, and a forceful effort by American diplomats to persuade Japan to buy more microelectronic goods from the U.S.A.

To deliver its warning that the Japanese were coming, the men of Silicon Valley selected one of their own—a senior statesman of the industry who seemed perfectly suited to the role. The designated spokesman was articulate, intelligent and immediately impressive. Even better, he was delighted to take on the job because the message was one with which he heartily agreed. He had an acute, almost paternal, fear for the future of his industry, and an angry conviction that the Japanese had not played fair. "Many of the Japanese practices are practices that we would see as unfair, illegal, or whatever," the spokesman told a congressional committee in the summer of 1983. ". . . It may be that we can hope the Japanese will play by the rules of our game, but I don't see any motivation for them to do so whatsoever, since they perceive that they are winning using the current strategy. And, indeed, they may win."

The spokesman was called to present the industry position week after week at hearings, seminars, conventions and press conferences all over the country. He performed the job as if he had been doing it all his life—but he had not. The spokesman had been a physicist, then an inventor, then a corporate manager and a venture capitalist—and had performed all those jobs with equal facility.

The spokesman was Robert N. Noyce.

11

The Patriarchs

Bob Noyce's metamorphosis from corporate manager to industry spokesman had come about as gradually, and as inevitably, as his earlier transition from inventor to manager. The change was not something he had planned; "I just sort of drifted into it," he says. But then if Noyce's life had turned out the way he originally planned, he would still be working in a physics lab somewhere, happily indulging his curiosity about the way things work and turning out monographs explicating interesting phenomena of solid-state physics.

In fact, Noyce's career had already begun to move out of the lab by January of 1959, when he hit upon The Monolithic Idea. That was his most important scientific discovery, and it was also, for all practical purposes, his last. There is something about Noyce—a fundamental confidence complemented by a compelling sense of presence—that prompts people to look to him for leadership. The physicists, chemists, and engineers working at Shockley Semiconductor in 1956 were required—it was part of William Shockley's unique management style—to give one another report cards. When the grades were tabulated, Noyce emerged as the consensus choice to be the group's technical director. Accordingly, when the "traitorous eight" left Shockley in 1957 to found Fairchild Semiconductor, the group turned to Noyce as soon as it became clear that somebody was going to have to act as a manager. His colleague Gordon Moore recalls that "Bob was everybody's choice. He was the natural leader."

In any case, it was a satisfying development, not least because it provided a whole new world of human endeavor to learn about. "Getting into management was just enormously exciting," Noyce recalled not long ago. "Because, first of all, I didn't know a damn thing about it—so that your learning rate goes up very, very rapidly. But secondly, management does become the focal point for all the information in the organization. Well, the guy who has the information has the power. . . . It's a very satisfying thing, particularly coming from a place where you're looking at a narrow field so you don't see the forest for the trees. And suddenly you're sitting in a balloon looking down from branch to branch and . . . for the first time you can see the whole."

More important, it just felt right—it fitted precisely with Noyce's evolving theory of corporate leadership—to have a technologist running a high-tech company. "One of the real problems with American business," he says, "is this notion that you can be trained in management, in some kind of generic form of management, and then you can manage any operation. But that absolutely does not work in a technical situation. The manager has to have an intuitive gut feel for what ought to be done in a particular situation, and if you don't have the technical background, if you haven't participated personally, you don't have that."

For the most part, the American semiconductor industry has been built and run by technologists, and to Noyce this is a key reason for its explosive success. "They say we were lucky," he explains. "Well, you can say you were lucky being up at bat in the World Series with two out in the ninth inning in the final game of the series and you happen to hit the home run. But the real point is that you have to be at bat at that time. Basically, you have to be in there participating and have the gut feel for what can be done."

Noyce's first management position at Fairchild Semiconductor was Director of Research and Development, a position considerably more important than it might sound because, for the first year or so, the firm was doing nothing but research and development. As the research developed into important products—the planar process and the integrated circuit—the upstart semiconductor division became one of the leading profit centers in the whole of Fairchild Camera and Instrument Corporation.

Fairchild's directors back East rewarded Noyce in the traditional ways: he was named General Manager of the division, then Vice President, which was basically the same job with a fancier title, then Group Vice President, which was even fancier. His salary began to climb toward the exalted levels reserved for athletes, entertainers and the top management of large corporations; but after 1959, when he and his seven cofounders sold their $500 worth of stock to the parent company for a quarter-million dollars each, salary was not so important a concern.

And yet, through the mid-1960s, Noyce was becoming more and more uncomfortable. Fairchild's directors wanted to run the new profit center their way, and this was something quite removed from what Bob Noyce and Gordon Moore had in mind. The two technologists were, as Noyce put it, "comfortable with risk," be it technical or financial. The generic managers back East at the home office were comfortable only with security, with the safe business play. Looking back later, Noyce could see that a fissure was inevitable. The break came in 1968, when Noyce and Moore decided—in sharp contrast to contemporary corporate wisdom—that money could be made by building computer memory circuits on a semiconductor chip. With the help of Silicon Valley's leading venture capitalist, Arthur Rock, the pair left Fairchild and founded Intel Corporation. They arranged things so that technologists would be in control. The firm's president was Robert Noyce; its chairman was Gordon Moore. Intel was to be governed, Noyce explained later, not by market surveys and financial analyses, but rather by "intuitive gut feel."

Noyce related all this one day in his office at Intel's headquarters, a sprawling rectangle of nondescript architecture alongside the Central Expressway in Santa Clara just shouting distance from an enormous amusement park named "Great America." To call Noyce's office an "office," though, is to stretch the language. The founder's work space, like the spaces allotted to everyone else at Intel, is really just a small cubicle bounded by four white metal partitions. Intel does not provide huge executive offices. Noyce's parking space in the lot outside is whatever space happens to be free when he arrives at work. Intel does not provide reserved parking for the brass. If Noyce arrives at the office later than 8:00 A.M., the official start of the day, his name goes on the late list like all other dawdlers'. Intel does not permit exceptions for those at the top. While junior executives of neighboring companies far less successful than Intel routinely drive to San Francisco for lavish expense-account lunches at exorbitant bistros, Bob Noyce's routine lunchroom is the Intel cafeteria. "The potato salad's pretty good," he says.

All of which is a direct reflection of the democratic, meritocratic, and studiously nonhierarchical management style that Noyce, Moore and another Fairchild refugee, Andrew Grove, custom-designed for their company. They wanted to nurture at Intel a feeling of "nobody here but us engineers"—a sense that the company is just a bunch of technologists working together to solve technical problems and keep two hops ahead of the competition. The thrust was to give each scientist, engineer and mathematician on the staff the ability—and the responsibility—to work at the leading edge of the industry. It is a management philosophy built for a company full of bright, curious, and driven superachievers—in short, for a company full of Noyces.

It is also a philosophy that works. "Intel became Silicon Valley's technology flagship," *The Wall Street Journal* reported, thanks to its pioneering development of semiconductor memory and the microprocessor. "Intel is still generally regarded as the most innovative American manufacturer of semiconductors," the *Journal* said in 1983, and Noyce is determined to see to it that the description fits for years to come. Intel has struggled through downturns with the rest of the industry, but on the whole it has experienced stupendous growth: fifteen years after it first opened for business with no product and a few dozen employees, Intel was a billion-dollar company with 20,000 people on the payroll.

Intel's corporate success was accompanied by extraordinary financial success for its founders. In 1969, at the age of forty-one, Noyce had become well-to-do with the quarter-million-dollar profit on his Fairchild stock. After the birth and rapid growth of Intel, though, he moved into the category of the seriously rich. On the day in 1971 when Intel's stock went public—at an opening price of $23.50 per share—Noyce's net worth went into the eight-digit range. Today his Intel holdings are about

1.5 million shares; the stock generally trades somewhere around $35 per share.

When a man who describes himself with the phrase "comfortable with risk" finds himself a multimillionaire, it is almost axiomatic that he is going to risk some of the money on financial ventures. For the past five years or so Noyce has been one of the more active venture capitalists in the high-tech sector. His portfolio today is studded with names like "Diasonics" and "Monocolonal Antibodies." He has also, naturally, put some money into the semiconductor and computer industries. As he cheerfully admits, his investment decisions have not always been the wisest. His wife asked him once about a start-up company down the road in Cupertino that was looking for investors. Noyce studied the prospects and warned her to stay away; no future there. The "no-future" company was Apple Computer, which went on to become the very symbol of financial success in the personal computer industry. (Fortunately, Mrs. Noyce had the good sense to ignore her husband's advice.) Bob himself put some money into a promising-looking outfit called Osborne Computer—only to see it lurch into bankruptcy two years later.

There have been, however, more Intels than Osbornes in Noyce's investment record, with the result that, by his fiftieth birthday, in 1983, the minister's son from Denmark, Iowa, was counting his assets in the tens of millions of dollars. He has used the money to pursue his unbounded appetite for new experiences. He flies his own jet to skiing vacations at his condominium in Aspen. He flies his own seaplane to boating vacations on the lakes of Northern California. Comfortable with risk at play as well as work, he has taken up gliding, hang gliding, paragliding, scuba diving, and assorted other sports, just for the hell of it. He still lives in the large but unspectacular house in Los Gatos where he and his first wife (he was divorced in 1974 and remarried a year later) raised their four children; but the house is now surrounded by lavish gardens—in his mid-forties, Noyce became intensely interested in gardening—accoutered with tennis court and swimming pool.

The combination of Noyce's vigorous atheltic life, his far-flung personal interests and his management work at Intel would have been sufficient to fill the days of most fifty-year-olds. For Noyce, though, the late 1970s found him looking beyond the low partition in his cubicle at Intel. Among much else, he started thinking about the future not only of his own company but of the entire semiconductor industry —and of American industry as a whole. When a group of Silicon Valley executives got together to form the Semiconductor Industry Association, Noyce was on the board. When journalists came to California to look into the chip business, Noyce's office became a regular stop. He was, in many ways, the perfect spokesman—a thoughtful, articulate founding father who could discuss both precise technical questions and broad issues of industrial policy.

By 1978, Noyce was spending as much time speaking at conferences, seminars, and Congressional hearings as he was at Intel. He stepped aside from day-to-day management at the firm—he now holds the title "vice chairman"—so that he could devote more time to industry-wide concerns. He became chairman of the Semiconductor Industry Association and a leader of its battle to stave off the challenge from Japan. He has become the elder statesman of Silicon Valley, the "scientist-cum-charmer," as a colleague put it—the official voice of his industry.

* * *

Twelve hundred miles to the southwest, in a cluttered office at the northern edge of Dallas, Jack Kilby, the quiet introvert, is happy to report that he is the official voice of no one but himself. He is perfectly willing to talk to the occasional reporter or historian who stops in, but for the most part he is tacitly and creatively engaged these days in the business he has always liked best: inventing.

For a while, in the years after he finished the Pocketronic calculator, Jack Kilby, too, seemed to be moving more or less inevitably into management. Eager to let him know that his groundbreaking work was appreciated, Texas Instruments rewarded Jack with a steady flow of raises, bonuses, stock options, and promotions. Promotion, for a man whose job was inventing, involved moving up to a management position supervising other inventors; eventually he became the No. 2 man in the hierarchy at T.I.'s research lab and development lab, and the top job there was easily within reach. By 1970, when he went to the White House to receive the National Medal of Science, he was one of the country's outstanding engineers; he seemed assured of more bonuses, more raises, and more promotions as long as he stayed at T.I.

And then, in November of 1970, he left.

Over the years, Jack had been thinking a great deal about the work he loved most—the demanding, creative, and ultimately rewarding job of inventing. It became more and more apparent to him that real creativity, artistic or technical, demanded real freedom—the kind of freedom that did not mesh with bureaucracies, whether governmental or corporate. "There is a basic incompatibility of the inventor and the large corporation," Jack wrote in a lecture on the subject. "Large companies have well-developed planning mechanisms which need to know at the beginning of a new project how much it will cost, how long it will take, and above all, what it's going to do. None of these answers may be apparent to the inventor." If Edison had worked for a big corporation, Kilby went on, there might have been no Edison light bulb, because the company's goals might not have matched the inventor's.

As he contemplated the work of Edison and Bell and other free-lance inventors—men who had defined the problems on their own and worked out their own solutions—Jack began to perceive that the grass might indeed be greener on the free-lance side of the fence. He had, as a matter of fact, enjoyed considerable freedom at Texas Instruments, but even there bureaucracy was beginning to chafe. His goal had always been to find nonobvious solutions to important problems; but anything nonobvious was anathema to corporate planners and accountants.

From his voluminous reading, Jack had put together a pair of lists comparing major inventions produced by large corporations with those made by individuals. The corporate inventors' list included, among other things, "Scotch" tape, television, nylon, and the integrated circuit. The individual inventor was credited with, among other things, air conditioning, penicillin, Xerography, and the zipper. Looking back on it today, Kilby is at a loss to explain any significant difference between the two lists. At the time, though, as he perused the two lists, they somehow seemed to prove a point. Today he realizes that the point had already been largely settled in his own mind. Free-lance was the way to go.

And so he packed up his books, his papers, and his favorite old slide rule and moved out to a place of his own in a low-rise office building on Royal Lane in North Dallas, about two miles down the LBJ Freeway from T.I. He's been there ever since, enjoying the satisfaction of working at his own pace on problems of his own choosing. "There's a certain amount of satisfaction," he says, "in setting your own goals, in being free to do what you decide is important, and not pursue somebody else's schedule. The freedom, that's what interests me about this."

He admits readily that the choice of freedom was not a particularly wise one in financial terms. "It was pretty damn close to stupid," he says. The economic rewards have been "pretty marginal."

The economic rewards for Jack Kilby have never approached the vast wealth accumulated by Robert Noyce, a fact that Kilby accepts philosophically. "Basically, engineers are hired by companies to do that kind of work," he says. "I don't get five percent of the value of everything that is ever sold, or anything of that sort . . . but I was rather well rewarded by T.I. for my work on the integrated circuit." His rewards at T.I. put his yearly salary well up into the five-figure range, an income that permitted Jack and Barbara and their two daughters to live in comfortable upper-middle-class style. It also permitted them to save enough to maintain their way of life during the lean years when Jack first set out on his own. Today Kilby draws an income for consulting work at Texas Instruments and as a faculty member at Texas A&M's Institute of Solid-State Electronics, a part-time appointment he received in 1978. But those jobs are essentially sidelines. Jack Kilby these days is just what he has wanted to be for a long time—an independent inventor.

The independent Kilby has received about a dozen patents, and they reflect the considerably wider scope of his ideas since he went out on his own. At his wife's suggestion, Jack started working on an "electronic intercept" device that keeps your telephone from ringing unless the call is one you want to take. Three years of tinkering resulted in Patent No. 3,955,354, "System for disabling incoming telephone calls." The gadget works perfectly but so far has been a dud in the market. Then there was Patent No. 4,001,947, "Teaching system," a small calculatorlike device that talks to a student as it teaches math, spelling, and other subjects. That idea, too, worked fine, but in this case Kilby was scooped by a competing product known as "Speak 'n Spell"—produced by Texas Instruments. The Kilby "Electronic check writer," Patent No. 3,920,979, has yet to earn its first dime, a description which also holds true for Patent No. 3,944,724, "Paging system with selectively actuable pocket printers."

For the first five years or so of his free-lance career, in other words, Kilby reveled in his new freedom, moving from one idea to the next as the spirit took him. What he needed, though—what he had always needed—was an important problem to work on. "There are a large number of real needs which the inventor can address," he said in his lecture on inventing. "The individual is free to choose a need that he thinks he may be able to satisfy. . . . *The definition of the problem becomes a major part of the innovation.*" (Kilby's emphasis.)

In the early 1970s, the United States and most of the rest of the world came up against a real need. Just as Kilby was setting up shop in his Royal Lane office, political and economic developments around the globe were defining a problem of unparalleled importance—a need that an electronic engineer familiar with semiconductor phenomena might be able to satisfy. By the early 1970s it became evident that the world could no longer depend on reliable sources of fossil fuels like oil and coal to generate electricity. There was a fundamental need for some other source of power—a source that was available to everyone on Earth, a source that was unlimited, a source that was free. In fact, this power source already existed—the sun. The problem that Jack Kilby took on in the mid-1970s, and has been working on ever since, was finding a cheap, efficient way to turn sunlight into electricity.

As it happens, this problem, like the Tyranny of Numbers, can be solved with semiconductors—particularly silicon. The basic idea here, known as the "Photovoltaic Effect," was one of the earliest discoveries of semiconductor physics. Physicists have known for more than a century that if you shine light on a strip of semiconductor material, such as silicon, electrons will start to move from one end of the strip to the other. The flow of electrons, as J. J. Thomson explained, is an electric current; thus if you hook up a pair of wires to the silicon strip, current will flow through the wires. That is, sunlight can generate electric current—the same current generated by the most expensive oil-, coal-, or nuclear-fueled power plants.

To make a practical solar generator requires a fairly large area of silicon. To produce the electricity an average house requires, for example, an area of silicon about the size of an average house's roof. A household solar generator, then, is going to consist of a rooftop lined with silicon—with wires to carry down the current induced by the sun. Most attempts to take this concept from paper to practice have involved putting fairly large slabs of silicon on the roof. Although this method seems brilliant in theory, the practical results have been disappointing.

Jack Kilby, looking over earlier work in this field, rejected the obvious answer—to use even larger slabs—and went the other way: the solar generator he has been working on diligently for the past five years or so uses tiny silicon pellets about the size of the head of a pin. The problem then became how to attach wires to the tiny pellets. After a long period of struggling with obvious answers, Jack came up with a nonobvious solution: eliminate the wires. That led to another problem, which forced Kilby to find another solution. His solutions to each new problem have been so promising that the government and various private companies—including Texas Instruments—have gotten interested. That corporate and government support has led to further progress.

For a man whose notion of heaven is to seize an interesting problem and solve it, Jack Kilby's lot is a happy one. He leads a quiet, thoughtful, and satisfying life, traveling from home to lab in his old Mercedes, thinking about his solar generator. His wife died not long ago, shortly after the couple's thirty-third anniversary. Although Jack sees his two daughters fairly often, and spends time with his colleagues from T.I., he leads a somewhat solitary life, thinking and reading, reading and thinking. "Jack is a thinker," says Willis Adcock, the sprightly engineer who first brought Kilby to Dallas a quarter-century ago to tackle the tyranny of numbers. "I would say Jack's got a good creative sense, he's got that, but the other thing that I liked when I hired him is that he is a persister. He just thinks a problem all the way through, works it through, and he doesn't stop until he's got it worked out. And you know, you can see the results."

You can see the results in tens of millions of homes around the world, in the thousands of companies and tens of thousands of new products The Monolithic Idea has spawned, in the awards, plaques, prizes, and citations that Jack Kilby and Robert Noyce have received from their peers.

The professional and technical awards have become so common, in fact, that for the most part neither Noyce nor Kilby pays much attention these days when word of a new one comes in the mail. But one day in 1982, Jack Kilby received an honor that really mattered to him, because it was proof he had succeeded at his chosen trade. He was inducted into the National Inventors' Hall of Fame, an august group of five dozen people—Edison, Bell, Ford, Shockley, the Wright brothers—whose inventions have made a difference.

On a sunny winter Sunday, a group of people gathered in the lobby of the Patent Office, just across the Potomac from the Washington Monument, for the Hall of Fame induction ceremony. Of the five inventors honored that year, only two—Kilby and Max Tishler, who started the vitamin industry in 1941 by synthesizing vitamin B_2—were still alive. Both were present. When the Secretary of Commerce called out Tishler's name, the aging chemist stood up and gave a long speech about how he got his idea and what it had meant. Then it was Jack Kilby's turn. He stood up for the briefest moment, looked around shyly at the audience, and quietly said, "Thank you." That was all. "He really didn't say a word during the whole thing," recalled Fred Ziesenheim, the Hall of Fame's president. "He just sat there like he was thinking about something. It looked like, no kidding, I sort of thought he was sitting there working out his next invention."

* * *

A scattering of newspapers around the country ran a brief story on the ceremony and on Jack Kilby's role as the patriarch of digital circuitry. A year later, when Robert Noyce was inducted into the same Hall of Fame for his part in the creation of the chip, a few papers again devoted a few inches of space to the subject. Every once in a while, some newspaper or business magazine reports on Noyce's role as the spokesman for Silicon Valley. But neither Noyce nor Kilby has ever received enough attention in the press to make their names familiar to any more than a minute fraction of his countrymen. Indeed, twenty-five years after they came up with the idea that changed the world and launched the microelectronics revolution—a revolution that has become a part of daily life for almost everyone on Earth—both Robert Noyce and Jack Kilby are cloaked in almost total obscurity.

It's a sign of the times. A few generations ago, men of this ilk—

men like Edison and Bell, Ford and Goodyear, whose inventions touched every life and spawned giant industries—were accorded enormous prominence. Although such things were not surveyed as carefully then as they are now, it seems fairly safe to assert that Edison was the best-known man in the country, and perhaps on Earth, within five years after he perfected the light bulb. The Wizard of Menlo Park, the "Napoleon of Science," he still ranked as the "Most Admired American" in a *New York Times* survey taken in 1922, when he was 75 years old and long finished with productive work. Alexander Graham Bell was a household name on the basis of his invention long before the nationwide "Bell System" was in place. Henry Ford and his Tin Lizzie became the stuff of myth, instantly recognized around the world as the symbols of the automobile age.

Today it seems clear that the chip will change the world as decisively as did the telephone and the automobile. The scientific and sociological savants all say so, and common sense tells us that they are right. But in the microelectronic age, Jack Kilby and Bob Noyce symbolize, if anything, only the modern lack of interest in the humans behind the machines. Not one American in ten thousand could name the two countrymen who invented the integrated circuit and launched the Second Industrial Revolution.

Is it because people really believe that computers contain "electronic brains"—and thus don't care to know about the human brainpower that made these mechanisms possible? Is it because we have swallowed the Orwellian notion that digital technology is a brutalizing, tyrannical force—and thus we don't want to honor, or even know, the men who made it? Is it because we have grown so accustomed to huge corporate and governmental enterprises that we no longer recognize individual invention? Is it because the media that purvey fame and recognition among our contemporaries—*People, Us,* "That's Incredible," "Good Morning America," and the like—don't trust their audience to appreciate genuine intellectual accomplishment?

The list of the "most admired" in today's world—a list assembled annually by the sophisticated surveying apparatus of the Gallup Poll—suggests that the current vogue in admiration runs heavily to political figures, with an occasional clergyman or entertainer thrown in. For three decades George Gallup and his organization have been asking Americans to name the two men they admire most. The answers vary little from year to year. The Pope and Billy Graham are generally on the list. Bob Hope and Walter Cronkite show up occasionally. The other names are drawn from government; Ronald Reagan, Edward Kennedy, and Henry Kissinger are among the hardy perennials in this garden. As the Gallup organization points out in a caveat accompanying its survey, the poll "tends to favor those who are currently in the news." It's hardly surprising, then, that men and women engaged in science and engineering tend to be left out, for such people are generally not treated as news—unless they become avid self-promoters (as Edison and Ford were), or unless, like William Shockley, they set aside their technical work and begin proselytizing for political causes.

And so, in an era when everybody is supposed to be famous for fifteen minutes, Jack Kilby and Bob Noyce have yet to come into their allotted quarter-hours. There are occasional stories about them in the newspapers and magazines, particularly in the local media of Dallas and Silicon Valley. Noyce has been written up here and there in the national press, as the progenitor of Silicon Valley and as a successful investor (*Time* magazine called him a "Financial Genie"). And on a fall day in 1983, Diane Sawyer flew to Dallas to interview Jack Kilby for the "CBS Morning News." The segment lasted about three hundred seconds, with Sawyer tossing out peppy questions and Jack responding in his slow, laconic way.

"I mean, if you have to think of one thing that kept the United States at the forefront of technology," Sawyer said, "it was really your invention." Kilby paused, stewing it over. "Well, I hadn't thought of it in those terms," he said quietly. "Have you made money from this invention?" Sawyer asked. "Some, yeah," Kilby replied. Things were just starting to get interesting when Sawyer got a signal from the director: time to move on. She turned quickly to the camera and said, "Coming up in a moment, Dr. Jerry Brodlie on how to handle the death of a pet." Jack Kilby's moment in the sun was over.

Brief as it was, that moment nicely captured the central irony. Our media-soaked society, with its insatiable appetite for important, or at least interesting, personalities, has somehow managed to overlook a pair of genuine national heroes—two Americans who had a good idea that has improved the daily lot of the world.

Author's Note

This book began with a disappearing typewriter.

On November 3, 1980—the day before Ronald Reagan won the presidency—I returned to my desk at *The Washington Post* after a year of constant travel covering the presidential campaign. To my distress, my cherished old typewriter had disappeared.

In its place, the *Post* had installed a computer terminal; from now on, I was to write my stories on that. At first, I resented this impostor. Quite soon, though, I came to realize that it was faster, quieter, easier to use, and far more efficient than the typewriter it had replaced. Within two weeks, I was a devout convert to the new technology; it had changed my daily life for the better.

One day I was mad at something and took a swat at the machine. Red lights blinked; beepers beeped; the screen went dark. A technician came over and opened the cabinet. "We'll have to replace a chip," she said. She pulled out a small black rectangle, maybe half an inch long, with a row of copper legs along each side—a plastic beetle—and dropped it into my palm. "This chip is the heart of the whole thing," she said.

Up to that moment, I had known in some abstract sense that the digital revolution was all based on a tiny "microchip." But now, that abstract awareness was translated into a tangible reality, nestled in the palm of my hand. I was determined to find out more about this chip and how it worked. Right then and there I dashed out to Reiter's technical bookstore on Pennsylvania Avenue to find a book on silicon chips. The salesman gave me the standard text—*Microelectronics,* by Professor Jacob Millman of Columbia University. There I found this sentence: "In 1958 Kilby conceived the Monolithic Idea, that is, the concept of building an entire circuit out of germanium or silicon." This hit me like a bolt of lightning. For the first time I realized the obvious: this miraculous chip was a man-made miracle. All the marvels of the computer age, all the "electronic brains" and "artificial intelligence," were simply products of the most powerful intelligence of all—the human brain.

I resolved to meet the humans responsible for this invention and find out what they had done and how they came to do it. This book is the result.

Many people helped me along the way, and my gratitude is enormous. My deepest thanks go to Jack Kilby and Robert Noyce, who were generous with their time and unfailingly patient with my dumb questions. Their colleagues Willis Adcock, Melvin Sharp, Mac Mims, Jerry Merryman, Norman Neureiter, Dick Perdue, Gordon Moore, Jean Hoerni, and Roger Borovoy went out of their way to help me, as did the librarians at both Texas Instruments and Intel Corp. Howard Warshaw of Atari Corporation took me on a guided tour through the inside of a chip.

The Science Reading Room at the Library of Congress is a great national treasure; the kind and knowledgeable staff there were the mainstays of my research from beginning to end. I am indebted as well to Morris Dertz and his reference staff at the Penrose Library of the University of Denver. The libraries of Princeton, Georgetown, and Stanford Universities and Manhattanville College also permitted me to use their resources. I owe thanks to the public libraries of Denver, Colorado, and Washington, D.C.—and particularly to the P. S. Miller branch of the Douglas County Library in the lovely foothills town of Castle Rock, Colorado. The Cleveland Institute of Electronics allowed me the use of educational materials. Professor Elizabeth Tuttle of the Physics and Engineering Department at the University of Denver and Jeff Singh of the Mathematics Department there provided valuable help.

Maralee Schwartz, the world's number one researcher, demonstrated time and again that no fact was too obscure for her to find. I once asked her to track down the speed at which an eyelid blinks; she was back in the wink of an eye with a detailed report. Valarie Thomas and James Schwartz also provided useful research assistance. Many U.S. government officials, including John McPhee of the Commerce Department, Isaac Fleischmann of the Patent Office, and Dr. Uta C. Merzbach of the Smithsonian Institution, were helpful along the way.

Several of my colleagues in Washington assisted with this book. The charming *bon vivant* Ben Bradlee and the cheery iconoclast Bill Greider both recognized early on that this was a great story and that I ought to pursue it. Howard Simons, Scott Custin, Ben Cason, Nick Lemann, and Joel Garreau provided thoughtful and useful advice. I am particularly indebted to Mary McGrory, Patrick Gross, Haynes Johnson, and Admiral H. G. Rickover for recognizing from the beginning that this effort would turn into something worthwhile. I'm grateful to Dr. William Leahy for his solid advice at the end. In New York, the wonderful Alice Mayhew proved to be an author's dream editor. Ann Godoff and Henry Ferris of Simon and Schuster also played an important role in shaping this book. Christopher McLehose of Collins, Ltd., provided wry and insightful advice. My agent, Rhoda Weyr, was my *fidus Achates* on the project from the first.

My fellow computer devotees may be interested to know that this book was written on a Z80-based Heathkit H-89 running CP/M and the Peachtext program; reference notes were maintained on an IBM-PC with PC-DOS and the PFS File/Report software.

My education in electronics and my first year of research on this book were supported by the Alicia Patterson Foundation. Without that estimable organization, there would be no book. I owe particular thanks to Alice Arlen, Joe Albright, Helen Coulson, and Cathy Trost at the foundation.

Last but foremost, Margaret Mary McMahon, McMahon Thomas Homer Reid, O'Gorman Katherine Penelope Reid, and Erin Andromache Wilhelmina Reid put up with me and the manuscript in cheery fashion, a task far more formidable than writing any book.

Washington, D.C.
1981

Castle Rock, Colorado
1984

A Note About Books

The nicest thing that happened to me in writing this book (other than being home every day for two years with my growing family) was meeting Jack Kilby and Robert Noyce. The *second* nicest thing, though, was discovering and reading all the fascinating books and articles that provided much of the factual information for the book you have just finished. Readers who plan to dig deeper into any of the events or characters of this story have a treat in store. However, if you stop in at a bookstore or library and just start browsing, you'll find there's much more in print than anyone could read. Accordingly, I have compiled the following road map to steer you toward some of the better books I ran across while writing this one.

The best nontechnical book I found on the general history of semiconductor electronics (although it is somewhat skimpy on the invention and development of the chip) is *Revolution in Miniature,* by Ernest Braun and Stuart McDonald (Cambridge University Press, 2nd ed., 1983). S. Handel, *The Electronic Revolution* (Penguin, 1967) is an easy and interesting discussion of electronics history from Benjamin Franklin to the invention of the transistor. The basic text on the "tyranny of numbers" was written by the man who coined that term: Jack A. Morton, *Organizing for Innovation* (McGraw–Hill, 1971). There are also two big, expensive coffee-table books on the history of electronics—The Editors of Electronics, *An Age of Innovation, The*

World of Electronics 1930–2000 (McGraw–Hill, 1981), which looks at the story through American eyes; and Elizabeth Antebi, *The Electronic Epoch* (Van Nostrand–Reinhold, 1982), which provides a European view of the same period.

Somewhat more technical, but still accessible to the lay reader, is Scientific American, *Microelectronics* (W. H. Freeman, 1977), which includes an overview of this technical revolution written by Robert Noyce. During the bicentennial year, several of the technical journals of the IEEE (the acronym is pronounced "I triple E" and stands for the Institute of Electrical and Electronic Engineers) gave a great deal of attention to technical history. The July 1976 edition of *IEEE Transactions on Electron Devices,* available at most large public or research libraries, contains first-person accounts of important inventions by many preeminent engineers, including William Shockley (on the transistor) and Jack Kilby (on the chip).

There are a number of reference works that focus on the matters covered in this book. I grew to rely on the Van Nostrand *Encyclopedia of Computer Science* (Van Nostrand 1976) and the McGraw–Hill *Encyclopedia of Science and Technology* (15 vols.; McGraw–Hill 1982). G. W. A. Dummer, *Electronic Inventions and Discoveries* (Pergamon Press, 3rd ed., 1983) is a listing of important developments. For information on individual scientists and engineers, Isaac Asimov, *Asimov's Biographical Encyclopedia of Science and Technology* (Doubleday, 2nd ed., 1982) is a delightful quick resource; the 16-volume *Dictionary of Scientific Biography* (Scribner's, 1970), edited by Prof. Charles C. Gillispie, is a classic piece of scholarship that provides clear and comprehensive biographies of hundreds of scientists and engineers (Kilby and Noyce are not included but surely will be in a later edition).

There are a slew of books about Thomas A. Edison. Two fairly recent biographies, Matthew Josephson, *Edison* (McGraw–Hill, 1959) and Robert Conot, *A Streak of Luck* (Seaview, 1979), are well done and rich with detail. The best source on Francis Upton's relationship with The Wizard of Menlo Park is Upton's own history, *Edison's Electric Light Bulb* (1881). For a pleasant history of physics in England in J. J. Thomson's time, there is Maurice Crowther, *The Cavendish Laboratory* (Science History Publications, 1950). J.J.'s son, the

Nobel laureate Sir George P. Thomson, wrote a fascinating memoir of his father, *J. J. Thomson and The Cavendish Laboratory in His Day* (Doubleday, 1965). The most delightful source of information on J. J. Thomson, though, is his modest, intriguing 1936 autobiography, *Recollections and Reflections,* which has been reprinted by Arno Press (1975). John A. Fleming wrote a memoir of his own, *Fifty Years of Electricity* (Marshall, Morgan, & Scott, 1921), and there is also a short remembrance by his devoted lab assistant, J. T. McGregor-Morris, *The Inventor of the Valve* (The Television Society, 1954). Georgette Carneal, *A Conqueror of Space* (H. Liveright, 1930), was an authorized biography of Lee deForest. Twenty years later, though, the inventor produced the autobiography with the immodest title mentioned in chapter 2 of this book: Lee deForest, *The Father of Radio* (Wilcox & Follett, 1950). C. P. Snow, *The Physicists* (Little, Brown, 1981), contains a warm portrait of Niels Bohr.

William Shockley's definitive *Electrons and Holes in Semiconductors, with Applications to Transistor Electronics* (Van Nostrand, 1950) is the basic text on semiconductor physics and on Shockley's way of thinking; for the first 100 pages or so, it is accessible to any diligent reader. The invention of the transistor has been related by the inventors in a series of journal articles and in the lectures they delivered upon receiving the Nobel Prize; the latter can be found in *Nobel Lectures—Physics, 1942–62* (Elsevier, 1964). A National Geographic book, *Those Inventive Americans* (National Geographic Society, 1971), provides a popularized look at the transistor and its fathers; a somewhat more technical presentation is offered in George L. Trigg, *Landmark Experiments in 20th-Century Physics* (Crane, Russak, 1975).

Except for this book, there is very little between covers about Jack Kilby and Robert Noyce. Gene Bylinsky, *The Innovation Millionaires* (Scribner's, 1976) includes a chapter on Noyce and Gordon Moore, but it deals more with financial than with technical accomplishments. Tom Wolfe wrote a terrific article about Noyce, "The Tinkerings of Robert Noyce," which appeared in the December 1983 issue of *Esquire.*

A reader who would like to delve deeper into patent law can obtain a number of interesting and instructive pamphlets from the Patent Office, Arlington, Va. 20231. I found *General Information Concerning Patents* (U.S. Department of Commerce, 1982) to be a good start.

For more detailed historical and legal information, the best one-volume source I have found is Peter D. Rosenberg, *Patent Law Fundamentals* (Clark Boardman, 1975).

The binary system and other mathematical principles underlying digital computer operations are discussed in the four-volume *The World of Mathematics* by James R. Newman (Simon & Schuster, 1956); volume 3 of this excellent set also includes a useful section on Boolean logic. Isaac Asimov's *Asimov on Numbers* (Doubleday, 1977) also discusses the binary system. The reader who is interested in math, though, will derive the most pleasure from the witty, insightful, and generally marvelous books of Eric Bell: two that pertain directly to material covered in this book are *Mathematics, Queen and Servant of Science* (McGraw-Hill, 1951) and *Men of Mathematics* (Simon & Schuster, 1937); the latter includes a fine portrait of George Boole. There is also interesting Booleana in Mary Everest Boole, *A Boolean Anthology* (Association of Teachers of Mathematics, 1972). Dover Press deserves our gratitude for keeping in print a paperback version of George Boole's masterpiece, *The Laws of Thought* (Dover Press, 1953). There is as yet no biography of Claude Shannon, but a reader might be interested in the book that launched the burgeoning field of "information theory"—that is, Claude E. Shannon, *The Mathematical Theory of Communication* (University of Illinois Press, 1949).

Computer history is just now emerging as an academic discipline of its own, and there will no doubt be some fine books written on the work of Von Neumann, Turing, and other computer pioneers. There is a good general history in Joseph C. Giarratano, *Foundations of Computer Technology* (Howard W. Sams, 1982). An important contribution to this literature is Herman Goldstine's *The Computer from Pascal to Von Neumann* (Princeton University Press, 1972), which is strangely organized but has the immediacy that could be conveyed by one who was present at the creation of the modern electronic computer. Andrew Hodges, *Alan Turing, The Enigma* (Simon & Schuster, 1983) and Steve J. Heims, *John von Neumann and Norbert Weiner* (M.I.T. Press, 1980) are the first complete biographies. Von Neumann's seminal paper "Preliminary Discussion of the Logical Design of an Electronic Computing Instrument" is reprinted in John Diebold, ed., *The World of the Computer* (Random House, 1973). A museum exhibit on computer history in 1973 gave rise to an interesting photographic history, Glen

Fleck, ed., *A Computer Perspective* (Harvard University Press, 1973). Jeremy Bernstein, *The Analytical Engine* (Morrow, 1981) is an enjoyable popular history of computers.

There are far more books than any one person could read on the inner workings of integrated circuits, microprocessors, calculators, and computers. The standard text on chips is Jacob Millman, *Microelectronics, Digital and Analog Circuits and Systems* (McGraw-Hill, 1979), which is excellent but can be heavy going for the non-engineer. At the other end of the scale is Larry Gonick, *The Cartoon Guide to Computer Science* (Barnes & Noble, 1983), which manages to be hilarious and quite informative at the same time; my eight-year-old and his forty-year-old father both learned from this book. Other useful texts were Bernard Grob, *Basic Electronics* (McGraw-Hill, 1984) and Jefferson Boyce, *Digital Computer Fundamentals* (Prentice-Hall, 1977).

The DIM-I calculator in this book is based to some degree on the "Simple As Possible" computer designed by Albert Paul Malvino in *Digital Computer Electronics, An Introduction to Microprocessors* (McGraw-Hill, 1983). Other books describing how a computer gets the answer include Gene McWorter, *Understanding Digital Electronics* (Texas Instruments Learning Center, 1978), Rodnay Zaks, *From Chips to Systems, An Introduction to Microprocessors* (Sybex, 1981), and the three-volume series by Adam Osborne, *An Introduction to Microcomputers* (Osborne/McGraw-Hill, 1982).

The flood of books on Japanese management that has poured forth upon this country in recent years includes several that discuss Japanese competition in semiconductor electronics. The most detailed picture of this phenomenon, though, emerges in a series of government studies, including *International Competition in Advanced Industrial Sectors: Trade and Development in the Semiconductor Industry,* U.S. Congress, Joint Economic Committee, Feb. 18, 1982; *U.S. Industrial Competitiveness,* U.S. Congress, Office of Technology Assessment, July 1981; *The Five-Year Outlook for Science and Technology 1981,* National Science Foundation (two vols.), 1981. The key source on that intriguing figure W. Edwards Deming is Dr. Deming's own text, *Quality, Productivity, and Competitive Position* (M.I.T. Press, 1982).

Notes

(*Fuller publishing information appears in the Note About Books.*)

Chapter 1

PAGE

9
honored in the textbooks
Cf. Millman, *Microelectronics,* p. xxii

9
Kilby only hoped . . . Kilby had been delighted . . .
some misgivings
Interview w/Kilby

9
reduced titanate capacitor
U.S. Patent No. 2,841,508 (1958)

9
steatite-packaged transistor
Kilby, "Transistor amplifier packaged in steatite," *Electronics,* Oct. 1956

10
It was that infatuation
Interview w/Kilby

10
the technical journals called it
Cf. Noyce, "Interconnections," *Proceedings of the IEEE,* Dec. 1964, p. 1648 ff.; Morton, "The Microelectronics Dilemma," *International Science & Technology,* July 1966, p. 38

10
the "neuristor"
Dummer, "Progress with extremely small elec. circuits," *New Scientist,* Feb. 7, 1963, pp. 283–84

10
A *Life* magazine reporter
Life, Nov. 5, 1956

11
was said to be working on
Interview w/Kilby

11
"He was enthused, but . . ."
Kilby, "Invention of the Integrated Circuit," *IEEE Transactions on Electron Devices,* July 1976, p. 650

11
"I was very . . ."
Interview w/Adcock

13
"I spent a lot . . ."
Interview w/Noyce

13
". . . it would be desirable . . ."
Noyce's lab notebook for Jan. 1959 (copy in author's file); also "The genesis of the integrated circuit," *IEEE Spectrum,* Aug. 1976, p. 47

13
"There was a tremendous . . ."
Interview w/Noyce

14
about 10,000 times per second
Cf. *Scientific American,* Jan. 1982, p. 124

14
known in England
Braun & MacDonald, *Revolution in Miniature,* p. 16

14
"debugging"
Giarratano, *Foundations of Computer Technology*

15
the first transistor radio
Cf. *Circuit News,* Apr. 15, 1979

16
"Interpretation of . . ."
Engineering Index, 1953

16
"Le Transistron . . ."
Ibid., 1955

16
"Circuito . . ."
Ibid.

16
"Tensoranalysis . . ."
Ibid.

16
"Perekhodnaya . . ."
Ibid., 1958

16
"Success story . . ."
Ibid., 1957

16
"Transistors Key . . ."
Ibid., 1953

16
"Méthodes d' . . ."
Ibid., 1962

16
"Fabulous . . ."
Ibid., 1953

16
"Switching . . ."
Ibid., 1962

16
"Electronic . . ."
Ibid., 1961

16
"Comment . . ."
Ibid., 1959

16
"Design . . ."
Ibid., 1960

18
had 350,000 electronic
Gilbert, ed., *Miniaturization,* p. 3

18
the Control Data CD 1604
Interview w/Kilby; *Electronics,* Apr. 1980, p. 81

18
"For some time now . . ."
Morton, *Proceedings of the IRE,* June 1958, p. 955 ff.

18
"Each element must . . .
Morton, *International Science & Technology,* July 1966, p. 38

19
Royal Radar Establishment
Braun & MacDonald, p. 108

19
"tend to exacerabate . . ."
Haggerty, in *Proceedings of the IEEE,* Dec. 1964, p. 1648

19
under the general title
Electronics News, Jan. 25, 1982, p. 16

19
"Operation Tinkertoy"
Kilby, in *Transactions on Electron Devices,* July 1976, p. 648

19–20
"In civilian equipment . . ."
Electronics, Oct. 1, 1957, p. 178

20
designers tried redundancy
Cf. Kilby, in *Electronics,* Aug. 7, 1959, p. 111

21
"After you become . . ."
Interview w/Noyce

22
"It was a situation . . ."
Ibid.

22
"The things that . . ."
Interview w/Kilby

23
"A large segment . . ."
Noyce, in *Scientific American,* Sept. 1977, p. 64

23
"There was just . . ."
Interview w/Kilby

23
"It was clear . . ."
Scientific American, Sept. 1977, p. 64

23
"The Second Industrial . . ."
National Academy of Science, *Microstructure Science, Engineering & Technology,* 1979, p. 1

23
"The most remarkable . . ."
In Forester, The *Microelectronics Revolution,* p. 1

23
"Integrated circuits are . . ."
Interview w/Noyce

Chapter 2

PAGE

24
observed for the first time in March 1883 . . .
Conot, *A Streak of Luck,* p. 337; Asimov, *Understanding Physics,* vol. 3, p. 42

25
he saw no future in it
Conot, p. 337 ff.; Asimov, vol. 3, p. 42

25
"the zero hour . . ."
G. Thomson, *J. J. Thomson,* p. 20

25
"Well, I'm not a scientist . . ."
Josephson, *Edison,* p. 283

26
"Dot" and "Dash"
Conot, p. 122

26
New York Daily Graphic
Ibid., p. 113

26
When Edison died . . .
Josephson, pp. 484–85

26
"The hair, beginning . . ."
Ibid., p. 132

26
"*Perseverentia omnia vincit* . . . 99 percent perspiration"
Conot

26
"at least we know 8,000 . . ."
Ibid.

27
"filled up with Latin . . ."; "ignorameter"
Josephson, pp. 440, 442

27
named him "Culture"
Conot, p. 133

28
being called the "Edison Effect"
Josephson, p. 278. Cf. Asimov, 3:42

28–29
"I have never . . . the Savanic world."
Josephson, p. 278; Conot, p. 337

29
we still talk today of "current" . . .
Handel, *The Electronic Revolution,* pp. 20–21

30
"that in a few years all the great . . ."
Crowther, *The Cavendish Laboratory,* p. 38

30
Trinity Prize; 1881 paper; succeeded in 1884
Crowther; G. Thomson, pp. 20–24;
J. J. Thomson, *Recollections & Reflections*

30
"Things have come to a pretty pass . . ."
Crowther, p. 109

30
"J. J. spent a good part of most days . . ."
G. Thomson, p. 114

30–31
In his memoirs . . .
J. J. Thomson, p. 435

31–32
"In the dusty lab'ratory,
'Mid the coils and wax and twine . . ."
G. Thomson, p. 98

33
"I have lately made . . ."; "corpuscles,"
Bragg and Porter, eds., *Royal Institute Library of Science,* pp. 43, 48

34
Fleming was determined . . . "There was a young fellow"
. . . "hooliganism."
McGregor-Morris, *The Inventor of the Valve,* pp. 62, 66

35
. . . came to be known as "crystal sets."
Cf. Asimov, vol. 3, p. 95

35
"I was pondering . . ."
Fleming, *Popular Radio,* Mar. 19, p. 175

36
"So nimble are . . ."
Fleming, *Fifty Years of Electricity,* p. 335

36
"absurd and deliberately . . ."
Tyne, *History of the Vacuum Tube*

37
"a more revolutionary step . . ."
Carneal, *Conqueror of Space,* p. 187

37
"Few inventions have had so many . . ."
Asimov, *Asimov's Biographical Dictionary of Science and Technology*

37
"Commonly known as . . ."
Electronics News, Jan. 25, 1982, Sect. 2, p. 10

38
"grid"; "Device for Amplifying . . ."
Asimov, vol. 3, p. 42; Carneal; Tyne

38
"iconoscope"
Cf. Asimov, vol. 3, p. 50

38–39
"radar"—crystal rectifiers
Cf. Handel, *The Electronic Revolution,* pp. 80–81

40
"the incarnation . . ." "He was not, as Einstein was . . ."
Snow, *The Physicists,* p. 51 ff.

41
"Contraria sunt complementa"
Biographical Encyclopedia of Scientists, p. 87

41
"In this picture we see a striking . . ."
Nobel Lectures—Physics, 1922–41, p. 8

44
"the vacant parking place . . ."
Shockley, *Electrons and Holes in Semiconductors,* p. 8

44
"N-type"; "P-type"
Cf. Millman, *Electronics,* pp. 12–13

45
"It has occurred to me today . . ."
Shockley, "The Path to the Conception of the Junction Transistor," *IEEE Transactions on Electron Devices,* July 1976, p. 603

45
"Try simplest cases"; "the will to think"; "In these four words . . ."
Shockley, "The Path . . .," pp. 600, 599–600.

45
"the wave function A (ϕ) for the hole-wave packet . . ."
Shockley, *Electrons and Holes . . .*, p. 444

46
taught a Freshman seminar . . .
"Brave New William Shockley," *Esquire,* Jan. 1973, p. 130 ff.

46
"THINKING about THINKING . . ."
IEEE Student Journal, Sept. 1968, pp. 11–16

46
"dysgenics"; "retrogressive evolution . . ."
Esquire, Jan. 1973. Cf. *New York Times,* May 3, 1970, p. 58, and Jan. 17, 1970, p. 30 (letter to ed. from Shockley)

46
"decline of our nation's human quality."
New York Times, Jan. 17, 1970, p. 30

46
"my research leads me inescapably . . ."
New York Times, Dec. 5, 1973, p. 38

46
a social policy to deal with . . .
New York Times, Dec. 13, 1973, p. 95

46–47
"Off Pig Shockley"; photograph; microphone went on the blink . . .
Esquire, Jan. 1973, p. 130 ff.

47
pursued the Republican nomination . . .
New York Times, Feb. 12, 1982, p. 16

47
"charming"
Interview w/Robert Noyce

47
"the vacuum tube and thermionics . . ."; "Don't worry, Walter . . ."
"Walter Houser Brattain," Bell Laboratories Record, Dec. 1972, p. 339 ff.

48
"one indeed needs to be . . ."
Nobel Lectures—Physics, 1942–62, Brattain, p. 377

48
"I feel strongly, however . . ."; "The thing I deplore . . ."
Bell Laboratories Record, Dec. 1972, p. 339

48
"if I wiggled it just right."
Trigg, *Landmark Experiments in 20th Century Physics,* p. 149

49
By 1946 . . . the operation of the semiconductor diode . . .
A History of Technology in the 20th Century, pt. 2, p. 1117

50
"Hearing speech amplified . . ."
Shockley, in *IEEE Transactions,* July 1976, p. 611

50
"My elation with the group's success . . ."
Ibid., p. 612

Chapter 3

PAGE

56
Quotations from Kilby
Interviews w/Kilby

56
60,000 patents every year
Fact sheet issued by U.S. Patent Office
(copy in author's file)

56
60,000 patents every year
"That's all right . . ."
Interview w/Kilby

57
"The definition of the problem . . ."; "Although invention is . . ."
Kilby, "The inventor's view," *Chemtech,* Mar. 1979, p. 66

57–58
"You only arrived . . ."; "You could design . . ."
Interviews w/Kilby

59–60
"they weren't going to . . ."; "It was sort of . . ."
Ibid.

61–62
Quotations from Kilby
Ibid.

62
Geophysical Research Corporation
"The Men Who Made T.I.," *Fortune,* Nov. 1961

63
"Our company now has two . . ."
Interview w/Adcock. Cf. *Fortune,* Nov. 1961

64
"discouraged"
Kilby, "Invention of the Integrated Circuit," *IEEE Transactions on Electron Devices,* July 1976, p. 650

64
"If Texas Instruments . . ."
Interview w/Kilby

65
"The following circuit elements . . ."
Kilby's lab notebook, July 24, 1958 (copy in author's file)

65
"Nobody would have made . . ."
Interview w/Kilby

65
Quotations from Kilby
Ibid.

66
"it was pretty damned cumbersome"
Interview w/Adcock

Chapter 4

PAGE

68–69
Quotations from Noyce
Interview w/Noyce

70
"They're high achievers. . . ."
Harvard Business Review, May/June 1980, p. 122

70
Quotations from Noyce
Interview w/Noyce

72
"It was simply . . ."
Ibid.

72
low-energy gas discharges
Journal of Applied Physics, Vol. 27, Aug. 1956, p. 843

72
dc transistor current amplification
Proceedings of the IRE, Sept. 1957, pp. 1228–43

72
"base widening punch-through"
Interview w/Noyce

73
semiconductor diffusion . . . 1000 degrees centigrade or so
Millman, p. 100

74
described Noyce as an introvert
The Economist, Dec. 27, 1980, p. 63

74–75
"There was a group . . ."
Interview w/Noyce

75
"the traitorous eight"
Hanson, *The New Alchemists,* p. 92; also interview w/Moore

75
"the realization . . ."
Interview w/Noyce

75
the return on his initial $500
Bylinsky, *The Innovation Millionaires,* p. 61 ff.

75–76
"Wall Street has spoken."
Financial World, Mar. 15, 1981, p. 30 ff.

76
Noyce's shares are worth
Time, Jan. 23, 1984, p. 48

76
"Here we were . . ."
Interview w/Noyce

76–77
Quotations from Noyce, Ralls
Ibid.

78
"all the bits and pieces . . ."
"The Genesis of the Integrated Circuit," *IEEE Spectrum,* Aug. 1976, p. 50

78
"In many applications . . ."
Noyce's lab notebook for Jan. 23, 1959 (copy in author's file)

78
a "unitary circuit structure . . ."
U.S. Patent No. 2,981,877, p. 1

Chapter 5

(Note: "Docket" refers to the file of documents submitted for Interference Proceeding No. 92,841, *Kilby* v. *Noyce,* on file at U.S. Patent Office, Arlington, Va.)

PAGE

79
The terrifying rumor
Testimony of Martin Fleit, July 1964—transcript in Docket. Also, interview w/Melvin Sharp

80
Mosher promised
Interview w/Sharp

80
"the right to exclude . . ."
U.S. Code—cited in *General Information Concerning Patents,* p. 2 (U.S. Dept. of Commerce, 1982)

80
"a written description . . ."
35 U.S.C. 112

80–81
Coca-Cola history
The Coca-Cola Co.—An Illustrated Profile (Coca-Cola Co., 1974); Chandler, *Asa Girggs Chandler* (Emory U.P., 1950), p. 119

81
trade secret protection of Coca-Cola formula
Rosenberg, *Patent Law Fundamentals,* pp. 15–16

81–82
Shredded Wheat Case
Kellogg Co. v. *National Biscuit Co.,* 305 U.S.C. 111 (1938)

82
"The sheets may be provided . . ."
General Information Concerning Patents, p. 17

82–83
"new"; "useful"
35 U.S.C. 101; no patent on nuclear weapons: 42 U.S.C. 2181

83–84
That first telephone patent
Rosenberg, p. 12

84
"Radically departing . . ." and other quotations from patent application
U.S. Patent No. 3,138,743

85
". . . shall set forth the best mode . . ."
35 U.S.C. 112

85
"integration" and "interconnection"
Interview w/Kilby; Kilby, "The Invention of the Integrated Circuit," *IEEE Trans. on Electron Devices,* July 1976

87
"flying wire picture"
Interview w/Roger Borovoy

87
"Instead of using . . ."
U.S. Patent No. 3,138,743

88
"We were still a brand new . . ."
Interview w/Noyce

89–90
Quotations from patent application
U.S. Patent No. 2,981,877

90
"Interference Proceeding"; "Board of . . ."
35 U.S.C. 13(a)

90
received a copy of . . . Form POL-102
In docket

91
"Motion to Dissolve . . ." and other documents
In docket

92
"Here I was . . ."
Interview w/Borovoy

92
"conducting material . . ."
U.S. Patent No. 3,138,743

92–93
experts' testimony
Testimony of Prof. G. Pearson (Stanford), Oct. 9, 1964; testimony of Prof. R. Maurer (Illinois), Nov. 24, 1964 (transcripts in docket)

93
"Request for . . ." and other motions
In docket

93
Borovoy's brief included; "Note that this . . ."
Copy of brief in docket

93
"we are not particularly . . ."
Opinion of the Patent Office Board of Patent Interferences, No. 92,841, Feb. 24, 1967 (in docket)

94
CCPA decision
Kilby v. *Noyce,* 416 F2d. 1391 (CCPA, 1969)

94
"Denied"
Kilby v. *Noyce,* cert. denied Oct. 12, 1970 (in docket)

94
in the summer of 1966
Interviews w/Sharp, Borovoy; *Electronics News,* May 9, 1966, p. 1

95
"Patent Appeals Court Finds . . ."
Cf. *Electronics News,* Nov. 24, 1969, p. 63

Chapter 6

PAGE

96
March 24, 1959; "Solid Circuits"
Interviews w/Kilby and Adcock. Cf. *Electronics,* Mar. 13, 1959

96
a lavish prediction . . . from T.I.'s president
Editors of Electronics, *An Age of Innovation,* p. 84

96
"It wasn't a sensation."
Interview w/Kilby

96
In its special issue
Electronics, Mar. 13, 1959

97
"match-head size solid-state circuit"
Electronics, Apr. 3, 1959, p. 11

97
"There was a lot of flack . . ."
Interview w/Kilby

97
one common line of analysis
Interview w/Noyce. Cf. Noyce, "Microelectronics," *Scientific American,* Sept. 1977, p. 67

97
the critics identified three; "These objections were . . ."
Kilby, "Invention of the Integrated Circuit," *IEEE Trans. on Electron Devices,* July 1976, p. 652

97
the giants of the industry . . . kept themselves clear
Ibid.; also Noyce, *Scientific American,* Sept. 1977, p. 68

98
"The synergy between a new component . . ."
Noyce, *Scientific American,* July 1977, p. 63

98
traditional economies were reversed
Ibid.

99
On Beyond Zebra
Seuss, *On Beyond Zebra* (Random House, 1955)

100–101
"positional notation"; invention of zero
Asimov, *Asimov on Numbers,* pp. 21–28

103
"ones-complement subtraction"
Malvino, *Digital Computer Electronics,* pp. 81–84

104–5
"ACCUSED HAD POWERFUL BRAIN"
Hodges, *Alan Turing: The Enigma,* p. 474

105
"One day ladies will take . . ."; "in about fifty years' time . . ."
Ibid., pp. 418, 417

105
The filmmaker Stanley Kubrick
Ibid., p. 533

105
Turing wrote, it was natural enough
Ibid., p. 320

105
"We feel strongly in favor . . ."
Von Neumann, *et al.*, "Preliminary Discussion of the Logical Design of an Electronic Computing Instrument," in Diebold, ed., *The World of the Computer,* p. 54

105
One was that the human race
Phillips, "Binary Calculation" (1936), in Randell, ed., *The Origins of Digital Computers,* p. 293

106
"The one disadvantage of the binary system . . ."
Diebold, pp. 55–56

106
"decoder"
McWhorter, *Understanding Digital Electronics,* pp. 8–12

107
"flash of psychological insight"
Mary E. Boole, *A Boolean Anthology,* pp. 61, 62; Kneale, "Boole and the Revival of Logic," *Mind,* vol. 57 (Apr. 1948), p. 152

107
"that his children not be allowed . . ."
Kneale, quoted in Newman, *The World of Mathematics,* vol. 3, p. 1854

108
"The design of the following treatise . . ."
G. Boole, *The Laws of Thought,* p. 1

108
"Sonnet to the Number Three"
Kneale, vol. 57

108
"No mere mathematician can understand . . ."
Mary Boole, p. 68

109
"$x(1-y)\ (1-z) + y(1-x)\ (1-z)$ . . ."
G. Boole, *The Laws of Thought*

109
$x = x^2$
G. Boole, p. 64

110
" 'You are sad,' the knight said . . ."
Carroll, *Through the Looking Glass*

110
In *Principia Mathematica*
Newman, vol. 3, pp. 1894–95

110
"the abstract doctrines . . ."
G. Boole

110
Vannevar Bush, had designed
Goldstine, *The Computer from Pascal to Von Neumann,* p. 90

111
"Programming a Computer for Playing Chess"
Philosophical Magazine, vol. 41 (1950), p. 256

111
"Squdgy fez, blank jimp crwth vox!"
New York Times, Aug. 16, 1972, p. 36

112
"It is possible to perform . . ."
Shannon, "A Symbolic Analysis of Relay and Switching Circuits," *AIEE Transactions,* vol. 57 (1938), p. 722

112
"In fact, any operation . . ."
Ibid.

114
"gates"
Cf. Millman, p. 123

Chapter 7

PAGE

116
a new line of six different . . . "micrologic elements"
Electronics, Mar. 31, 1961, p. 91

116
a similar series of "solid circuits"
Interview w/Kilby; *An Age of Innovation,* p. 83

116
priced at $120
Electronics, Mar. 31, 1961, p. 76

116
"There was the natural . . ."
Interview w/Noyce

117
"learning curve"
Braun and MacDonald, p. 82

118
"The space program badly needed . . ."
Interview w/Kilby

118
U.S. electronics companies frequently complain
Cf. "Competitive Factors Influencing World Trade in Semiconductors," Hearings before House Subcommittee on Trade, No. 96–62, Nov. 30, 1979

118
A study published in 1977
Linville and Hogan in *Science,* Mar. 18, 1977, p. 1109

118
federal government remained the largest buyer
Forbes, Feb. 15, 1971, p. 21; Braun and MacDonald, pp. 141–42

119
"The general rule of thumb . . ."
Interview w/Noyce

119
specifications called for every single component
Interview w/Adcock; Braun & MacDonald, p. 114

119
testing, retesting, and re-retesting
J. Morton, *Organizing for Innovation,* p. 106; *Business Week,* Apr. 14, 1962, p. 168

119
services went off in three different directions
Kilby, "Invention of the Integrated Circuit," *IEEE Trans. on Electron Devices,* July 1976, p. 649

119
"molecular electronics"
Cf. Air Force Systems Command, *Integrated Circuits Come of Age,* p. 4 (U.S.A.F., 1966)

119
"The idea of it was . . ."
Interview w/Noyce

120
"T.I. has always followed a strategy . . ."
Interview w/Kilby

120
"The development of integrated circuits . . ."
Integrated Circuits Come of Age, p. 1

120–121
"Almost an insult"; "The missile program and the . . ."
Interview w/Noyce

121
The designers of Minuteman II . . . 4,000 chips per month
Integrated Circuits Come of Age, p. 16; *An Age of Innovation,* p. 325

121
About 500,000 . . . were sold . . . and quadrupled again
Kraus, *An Economic Study of the U.S. Semiconductor Industry,* pp. 80, 209

121–22
"From a marketing standpoint . . ."
Interview w/Kilby

122
The first chip sold for the commercial market
Integrated Circuits Come of Age, p. 20

122
In 1963 the price of an average chip . . . $1.23
"A Report on the U.S. Semiconductor Industry," U.S. Commerce Dept. 99, p. 50; Electronic Industries Association, *Market Data Book 1975,* p. 84

123
"A single speck of dust . . ."
Interview w/Noyce

123
"I just did it sort of tongue-in-cheek. . . ."
Interview w/Moore

124
SSI, MSI, LSI, VLSI, ULSI
Millman, p. xxiv

124–25
"We would cruise comfortably . . ."
G. Moore, "The Cost Structure of the Semiconductor Industry," *IEEE Transactions Consumer Electronics,* Feb. 1977

125
"Progress has been astonishing . . ."
Noyce, "Microelectronics," *Scientific American,* Sept. 1977, p. 65

126
IBM brought out . . . System 360 . . . competitors had to
An Age of Innovation, p. 339 ff.

127
A monograph that appeared
Hodges, "Large-Capacity Semiconductor Memory," *IEEE Proceedings,* July 1968, p. 1148

128
Intel rang up a grand total
Fortune, Nov. 1973, p. 142

129
"linear" or "analog" integrated circuit
Millman, p. 333

129
"We reached a point . . ."
Interview w/Moore

129–30
To maintain its explosive rate of growth
Cf. Moore, "The Cost Structure . . ."

Chapter 8

PAGE

131
"Pervasiveness"
Interviews w/Kilby and Adcock; Noyce, "Microelectronics," *Scientific American,* Sept. 1977, p. 69

132
"We knew we were doing pretty well . . ."
Haggerty, "A New Challenge," remarks at 25th Anniversary Observance, Transistor Radio (Texas Instruments, 1980), p. 1

132
Some Texas Instruments people say
Interviews w/Adcock, Merryman

132
"if that little outfit down in Texas . . ."
Haggerty, "A New Challenge," p. 1

133
Haggerty started talking
Interview w/Kilby

133
"I sort of defined . . ."
Ibid.

134
"It was a miserable . . ."; "He's one of these guys . . ."
Ibid.

134
is going to find an answer if you think
Interview w/Merryman

135–36
"The basic rule was . . ."; "That way, all the bits . . ."
Ibid.

136
"We were trying . . ."
Ibid.

137
"The thing was all laid out . . ."
Ibid.

137–38
"How many housewives . . ."
Braun and MacDonald, p. 175

138
"the first piano in history . . ."
McDonald, "New Features Expand Calculator Market," *Leisure Time Electronics,* Aug. 1981, p. 37

138
"with a warm, synthesized voice"
"Texas Instruments Electronic Calculators 1982," p. 16

139
the first digital watch
An Age of Innovation, p. 398

140
The story of the microprocessor begins
Noyce and Hoff, "A History of Microprocessor Development at Intel," *IEEE Micro,* Feb. 1981, p. 8 ff; Libes, "The First Ten Years of American Computing," *Byte,* July 1978, p. 64 ff.

141
"If this continued . . ."
Noyce and Hoff, Feb. 1981, p. 8

141
Accordingly, Busicom told Intel
Ibid., p. 13

142
"insert intelligence into many products . . ."
Ibid.

142
"computer on a chip"
Cf. advertisement in *Electronics News,* Nov. 15, 1971

143
the first patent awarded for a microprocessor
U.S. Patent No. 3,757,306

143
actually was a computer on a chip
U.S. Patent No. 4,074,351

143
The introductory price was $200
Libes, July 1978, p. 68

144
"Project Breakthrough! . . ."
Popular Electronics, Jan. 1975

144
"Homebrew Computer Club"
Time, Jan. 3, 1983

145
"Clearly, a world with . . ."
Interview w/Noyce

Chapter 9

PAGE

146
a summer day in 1976
New York Times, July 11, 1976, sec. 3, p. 13; interview w/Dr. Uta Mersback of Smithsonian

146
had watched sales fall
New York Times, Jan. 3, 1982, sec. 4, p. 9

146
"Calculator usage is now . . ."
New York Times, July 11, 1976, sec. 3, p. 13

147
"That silent computational . . ."
Petroski, Reflections on a Slide Rule," *Technology Review,* Feb./Mar. 1981, p. 32

147
"The absence of a decimal point . . ."
Ibid., p. 35

147
"It has a sort of . . ."
Interview w/Kilby

150
TMS 1000C
"Semiconductor Products Master Selection Guide," p. 9, Texas Instruments, 1982

151
This is a fairly standard clock rate
McWhorter, "The Small Electronic Calculator," *Scientific American,* March 1976, p. 88

153
"propogation delay"
Cf. Millman, pp. 161, 225

158
"LED" and "LCD"
Texas Instruments, *Understanding Digital Electronics,* pp. 1–7

161
"truth table"
Scientific American, *Microelectronics,* pp. 29–30

Chapter 10

PAGE

164
On the morning of June 27
Interview w/Kilby

164
"TO ALL TO WHOM . . ."
U.S. Patent Office, Letters Patent No. 3,819,921

165
Since its first appearance in the 1930s
The Impact of Electronics on the U.S. Calculator Industry 1965 to 1974, U.S. Dept. of Commerce, Nov. 1975, pp. 6–8

165–66
"With innovative products . . ."; "an excellent illustration . . ."
Ibid., pp. 3, 1

166
T.I. sued Casio
Interview w/Melvin Sharp, patent counsel, T.I.

166
"Since 1974, the situation has once again . . ."
The U.S. Calculator Industry Since 1974, U.S. Dept. of Commerce, May 1977, p. 1

166
Japanese made about 45 percent
Ibid., p. 7

166
By the early 1980s
Interview w/John McPhee, Office of Business Research and Analysis, U.S. Dept. of Commerce

167
"This pattern of market activity . . ."
The U.S. Calculator Industry Since 1974, p. 1

167
television history
Dummer, *Electronic Inventions and Discoveries,* 3rd ed., p. 27; Bingley, "A half-century of television reception," *IRE Proceedings,* May 1962, pp. 799–802; Handel, pp. 68–72, 129 ff.

167
the term "television"
Antebi, *The Electronic Epoch,* p. 149

168
Zworykin-Farnsworth
Handel, pp. 68–72

168
6,000 television sets . . . two years later, 7 million
Baranson, *The Japanese Challenge to U.S. Industry,* p. 75

168
U.S. television manufacturers developed a marketing
U.S. Industrial Competitiveness, U.S. Congress, Office of Technology Assessment, 1981, p. 78 ("OTA Study")

169
purchased the rights to more than 400 patents
Baranson, p. 38

169
The Japanese television firms approached
OTA Study, pp. 78–9; *The Five-Year Outlook for Science and Technology,* National Science Foundation, vol. 2 (1981), pp. 409–10

169
The results
Baranson, pp. 75–78; *Five-Year Outlook,* p. 409–11

169
But the color TV market turned into
Baranson, p. 77

170
Admiral; Motorola
Ibid., table 5–1, p. 77

170
Some Japanese competitors were "dumping" . . . ;
"Orderly Marketing Agreement"
Ibid., p. 82; OTA Study, pp. 114–16

170
to a single chart
The U.S. Consumer Electronic Industry and Foreign Competition, U.S. Dept. of Commerce, May 1980, p. 2

171
between 1964 and 1970, royalty payments
Tilton, *International Diffusion of Technology,* p. 166

171
For some U.S. firms, particularly Fairchild
Finan, *The International Transfer of Semiconductor Technology Through U.S.–based Firms,* National Bureau of Economic Research, 1975, p. 47; Interview w/Borovoy, Intel Corp.

171
"American firms have generally . . ."
Tilton, p. 166

171–72
T.I. spurned royalty payments . . . The Dallas firm was . . .
Ibid., pp. 146–47; *Washington Post,* Mar. 23, 1980, p. E2

172
MITI's "Vision" for the 1980s
International Competition in Advanced Industrial Sectors: Trade and Development in the Semiconductor Industry, U.S. Congress, Joint Econ. Comm., Feb. 18, 1982, p. 56

172
In the 1950s, for example . . . Japan's foray into
Interview w/Kobayashi; OTA Study, pp. 78–92

172–73
Just as MITI planned it
OTA Study, pp. 79–94

173–74
By 1980 the Japanese had 42 percent . . . by 1983 they
Wall Street Journal, Nov. 17, 1981, p. 1

174
"The television people woke up . . ."
New York Times, Jan. 18, 1978, p. D1

175
"I just don't want to pretend . . ."
New York Times, April. 4, 1982, sec. F, p. 14

175
The dissenters pointed out
Washington Post, Mar. 23, 1980, p. E2;
Inquiry, Nov. 1983, p. 12

176
"The Anderson Bombshell"
The Rosen Electronics Letter, Mar. 31, 1980, p. 3

176–77
Anderson paper
Printed in *Market Conditions and International Trade in Semiconductors,* No. 96–60, House Committee on Ways and Means, Apr. 28, 1980, pp. 36–41

177
"U.S. Microelectronics Firms . . ."
Wall Street Journal, Nov. 17, 1981, p. 20

178
"Well, I'm not doing anything . . ."
Interview w/Deming

179
"14 points"
Deming, *Quality, Productivity, and Competitive Position,* pp. 11–17

179
"It's so simple . . ."; "By describing statistically . . ."
Deming, lecture at Falmouth, Mass., Mar. 25, 1983

180
"To save 15 cents per spool . . ."; ". . . before it gets to your inspectors"
Interview w/Deming

180
"The total cost to produce and dispose of . . ."
Deming, *Quality, Productivity* . . . , p. 21

181
"Brilliant applications burned, sputtered . . ."
Ibid., p. 101

181
"I predicted . . ."; "Once you convince . . ."
Interview w/Deming

182
"Third Wave"
King, "Introducing the Deming Approach to Management Growth Opportunity Alliance," Lawrence, Mass., 1983, p. 1

182–83
"The way to compete . . ."
Interview w/Deming

183
the SIA's shopping list
"The International Microelectronic Challenge," Semiconductor Industry Association, May 1982, pp. 21–32, 41

184
"Many of the Japanese practices are . . ."
Field Hearing on International Policy, House Comm. on Banking, Finance, and Urban Affairs, Aug. 18, 1983, p. 21 (stenographic transcript)

Chapter 11

PAGE

185
"I just sort of drifted into it."
Interview w/Noyce

186
"Bob was everybody's choice"
Interview w/Moore

186
"One of the real problems . . . for what can be done."
Interview w/Noyce

187
when he and his seven cofounders sold
Bylinsky, p. 64

187
"comfortable with risk"; "intuitive gut feel"
Interview w/Noyce

188
"Intel became . . ."; "Intel is still . . ."
Wall Street Journal, Feb. 4, 1983, p. 21

188
at an opening price of . . .
Ibid., Oct. 14, 1971, p. 25

188–89
Today his Intel holdings . . .
Time, Jan. 23, 1984, p. 48

189
His wife asked him once
Field Hearing on International Policy, Subcommittee on Trade, House Committee on Banking, Finance, and Urban Affairs, Aug. 18, 1983, p. 26

190
"scientist-cum-charmer"
Jerry Sanders, quoted in *Wall Street Journal,* Feb. 4, 1983, p. 1

191
"There is a basic . . ."
Kilby, "The Inventor's View," *Chemtech,* Mar. 1979, p. 66

192
"There's a certain amount . . ."
Interview w/Kilby

192
"It was pretty damn close . . ."
Ibid.

192
"Basically, engineers are hired . . ."
"CBS Morning News," Oct. 24, 1983

193
"There are a large number . . ."
Chemtech, Mar. 1979, p. 66

194
"I would say Jack's . . ."
Interview w/ Adcock

195
"He didn't really say a word . . ."
Interview w/ Zeisenheim

197
Sawyer-Kilby interview; "Coming up in a moment . . ."
"CBS Morning News," Oct. 4, 1983 (transcript in author's files)

Index